Native Warm-Season Grasses:
Research Trends and Issues

WITHDRAWN

Native Warm-Season Grasses: Research Trends and Issues

This special publication is based on technical papers presented at the Native Warm-Season Grass Conference and Expo that took place in Des Moines, IA, 12 to 13 Sept. 1996.

Editors
Kenneth J. Moore and Bruce E. Anderson

Editor-in-Chief CSSA
Jeffrey J. Volenec

Managing Editor
David M. Kral

Associate Editor
Marian K. Viney

CSSA Special Publication Number 30

Crop Science Society of America
American Society of Agronomy
Madison, Wisconsin
2000

Cover design adapted from Fig. 4–1
(Mitchell & Moser, 2000, this publication)

Copyright © 2000 by the Crop Science Society of America
American Society of Agronomy

ALL RIGHTS RESERVED UNDER THE U.S. COPYRIGHT ACT OF 1976
(PL. 94-553).

Any and all uses beyond the limitations of the "fair use" provision of the law require written permission from the publisher(s) and/or the author(s); not applicable to contributions prepared by officers or employees of the U.S. Government as part of their official duties.

Crop Science Society of America, Inc.
American Society of Agronomy, Inc.
677 South Segoe Road, Madison, WI 53711 USA

The views expressed in this publication represent those of the individual Editors and Authors. These views do not necessarily reflect endorsement by the Publisher(s). In addition, trade names are sometimes mentioned in this publication. No endorsement of these products by the Publisher(s) is intended, nor is any criticism implied of similar products not mentioned.

Library of Congress Registration Number: 99 085865

Printed in the United States of America

NORTH DAKOTA
STATE UNIVERSITY

MAR 2 7 2001

SERIALS DEPT.
LIBRARY

37-0761
Astell, Ann W. **Political allegory in late medieval England.** Cornell, 1999. 218p bibl index afp ISBN 0-8014-3560-9, $35.00

In this intertextual study of literature and history of 14th- and 15th-century England, Astell (Purdue Univ.) examines allegorical practice, intended audiences, and the political import of major works by William Langland, John Gower, Geoffrey Chaucer, the Gawain poet, and (to a lesser extent) Thomas Malory. The introduction and first chapter explain the rhetorical and poetic theory of medieval allegory in classical and Augustinian terms. This discussion establishes the ground for five subsequent chapters, in which the author establishes literary dates and associates specific historical contexts with multiple literary meanings: e.g., John Ball's letters and the Peasant's Revolt decode *Piers Plowman*. Astell compares a segment of *Confessio Amantis* and portions of Chaucer's works with events in Ricardian history and precisely dates and interprets *Gawain and the Green Knight* by events of 1397-99. In the typological allegory of *Morte D'Arthur*, Astell associates Malory's adulterous Guenevere with Joan of Arc and Margaret of Anjou and matches Guenevere's salvation with Joan's rehabilitation. Included are a preface, conclusion, and notes. Learned and thorough, this book will interest scholars of early modern as well as late medieval England.—*F. K. Barasch, emeritus, Bernard M. Baruch College, CUNY*

CONTENTS

Foreword	vii
Preface	ix
Contributors	xi
Conversion Factors for SI and non-SI units	xiii

    Introduction
        Martin A. Massengale 1

1. Escape and Rumen Degradable Protein Fractions in Warm-Season Grasses
    Daren D. Redfearn and Karla Jenkins 3

2. Fiber Composition and Digestion of Warm-Season Grasses
    Kenneth J. Moore and Dwayne R. Buxton 23

3. Morphology of Germinating and Emerging Warm-Season Grass Seedlings
    Lowell E. Moser 35

4. Developmental Morphology and Tiller Dynamics of Warm-Season Grass Swards
    Rob Mitchell and Lowell E. Moser 49

5. Growing Legumes in Mixtures with Warm-Season Grasses
    J. Ronald George, Kevin M. Blanchet, and Randy M. Gettle ... 67

6. Improving Warm-Season Forage Grasses Using Selection, Breeding, and Biotechnology
    Kenneth P. Vogel 83

7. An Overview of Seed Dormancy in Native Warm-Season Grasses
    Allen D. Knapp 107

8. Genetic Advances in Eastern Gamagrass Seed Production
    Chet Dewald and Bryan Kindiger 123

9. Cutting Management of Native Warm-Season Grasses: Morphological and Physiological Responses
    Matt A. Sanderson 133

10. Use of Warm-Season Grasses by Grazing Animals
    Bruce E. Anderson 147

11 Managing Weeds to Establish and Maintain Warm-Season Grasses
    Rob Mitchell and Carlton Britton 159

12 Fertilization of Native Warm-Season Grasses
    John J. Brejda 177

FOREWORD

Native warm season grasses of the continental USA represent major sources of biological diversity (more than 400 species) from a number of ecosystems. The potential is very good for matching native species to grazing and harvested forage needs in site specific climatic and edaphic situations. There are species reestablishment and management challenges, however, for both researchers and forage–livestock managers for widespread adoption and use.

Warm season grasses are erratic seed producers and their seeds typically exhibit various types of dormancy that limit uniform and rapid germination. Seed sizes vary from the very small sand dropseed (*Sporobolis cryptandrus*: 12 300 seeds g^{-1}) to the relatively large burs of buffalo grass (*Buchloe dactyloides*: 110 seeds g^{-1}). Seed size variation and low seedling vigor of some species creates seeding depth limitations and stand establishment problems. The requirement of 20 to 30°C temperature for germination and the low seedling vigor contribute to weed competition during establishment. Growth patterns of species differ and may lead to improper harvesting or grazing techniques creating openings for weeds to invade or compete. Thus managing native warm season species for quality forage and stand persistence requires an understanding of their changing nutritional value and growth patterns.

Improving and reestablishing native warm season grasses for both extensive and intensive management systems requires the collaborative efforts of a diverse group of researchers combining their expertise with ecosystems managers, i.e., farmers and ranchers, to overcome genetic and management obstacles in order to optimize the use of these valuable native species.

We think you will see both the challenges and the opportunities that native warm season grasses present as you read the chapters assembled in this special publication.

Vernon B. Cardwell
ASA President

Ronald L Phillips
CSSA President

PREFACE

Interest in native warm-season grasses is growing in the Midwest and Great Plains as researchers and producers look for alternative species to meet specific management objectives. Many of the pasture species commonly grown in this region are cool-season species that originated in Europe and Asia. Several of these have been adapted and become naturalized; however, the growth of many introduced species during the summer is limited by reduced photosynthetic rates at warmer temperatures. Because of this, their productivity is limited during that period of the growing season when solar radiation is most available. The uneven seasonal distribution of forage production from cool-season species is one of the primary factors complicating pasture management. Native plant communities are relatively more efficient at capturing solar radiation. Biomass accumulation in native prairies, which are dominated by warm-season grasses, essentially follows the same pattern as the solar radiation curve. Therefore, from a seasonal growth perspective, including native warm-season grasses in pasture systems makes good ecological sense.

There are problems with using warm-season grasses as pasture, however. Native warm-season grasses generally have a lower nutritive value than cool-season grasses. Most of the native warm-season grasses are climax species that evolved under intermittent defoliation by grazing herbivores and periodic burning. Because of this, they do not persist well under continuous grazing pressure and are subject to invasion by less desirable species unless prescribed burning or other weed management strategies are followed. These and other issues pertaining to the production and use of native warm-season grasses are discussed in the chapters that follow.

This special publication is based on technical papers presented at the Native Warm-Season Grass Conference and Expo that took place in Des Moines, IA 12 to 13 Sept. 1996. The objectives of the conference and this book are to present current research findings and discuss issues related to the production and use of native warm-season grasses.

The editors wish to express their gratitude to the symposium planning committee who developed the original outline for this publication, Dr. Martin A. Massengale for writing the Introduction, and to the authors for their efforts in the preparation of the chapters.

Kenneth J. Moore
Iowa State University, Ames

Bruce E. Anderson
University of Nebraska, Lincoln

CONTRIBUTORS

Bruce E. Anderson	Professor, Department of Agronomy, University of Nebraska, 353 Keim Hall, Lincoln, NE 68583-0910; 402/472-6237
Kevin M. Blanchet	12355 220th St. East, Hastings, MN 55033; 651/480-7739
John J. Brejda	Research Agronomist, USDA-ARS Wheat, Sorghum, and Forage Research Unit, University of Nebraska, 344 Keim Hall, Lincoln, NE 68583-0939; 402/472-1566
Carlton M. Britton	Professor, Department of Range, Wildlife, and Fisheries Management, Texas Tech University, Box 42125, Lubbock, TX 79409-2125; 806/742-2842
Dwayne R. Buxton	National Program Leader, USDA-ARS, 5601 Sunnyside Ave., Room 4-2210, Beltsville, MD 20705-5139; 301/504-6725
C. L. Dewald	Research Agronomist, USDA-ARS, 2000t S. 18th St., Woodward, OK 73801; 405/256-7449
J. Ronald George	Agronomy Department, Iowa State University, 1202 Agronomy Hall, Ames, IA 50011; 515/294-1360
Randy M. Gettle	Asgrow Seed Company, 310 Main St. NE, P.O. Box 447, Mapleton, MN 56065; 507/524-3475
Karla Jenkins	Assistant Professor, Department of Animal Science, University of Tennessee, P.O. Box 1071, Knoxville, TN 37901; 615/974-3456
Bryan Kindiger	Research Geneticist, USDA-ARS, Grazinglands Research Laboratory, 2707 West Cheyenne St., El Reno, OK 73036; 405/262-5291
Allen D. Knapp	Associate Professor, Agronomy Department, Iowa State University, 1569 Agronomy Hall, Ames, IA 50011-1010; 515/294-9830
Martin A. Massengale	Center for Grassland Studies, 222 Keim Hall, P.O. Box 830953, University of Nebraska, Lincoln, NE 68583; 402/472-4101
Rob Mitchell	Assistant Professor, Department of Range, Wildlife, and Fisheries Management, Texas Tech University, Box 42125, Lubbock, TX 79409-2125; 806/742-2842
Kenneth J. Moore	Professor, Department of Agronomy, Iowa State University, 1567 Agronomy Hall, Ames, IA 50011; 515/294-5482
Lowell E. Moser	Professor, Agronomy Department, University of Nebraska, 352 Keim Hall, P.O. Box 830915, Lincoln, NE 68583-0915; 402/472-1558

Daren D. Redfearn — Plant Physiologist and Assistant Professor, USDA-ARS, Louisiana State University Agricultural Center, Southeast Research Station, Franklinton, LA 70438; 504/839-3740

Matt A. Sanderson — Research Agronomist, USDA-ARS Pasture Systems and Watershed Management Research Laboratory, Building 3702, Curtin Road, University Park, PA 16802; 814/865-1067

Kenneth P. Vogel — Research Geneticist and Research Leader, USDA-ARS, University of Nebraska, 344 Keim Hall, P.O. Box 830937, Lincoln, NE 68583-0937; 402/472-1564

Conversion Factors for SI and non-SI Units

Conversion Factors for SI and non-SI Units

To convert Column 1 into Column 2, multiply by	Column 1 SI Unit	Column 2 non-SI Units	To convert Column 2 into Column 1, multiply by
Length			
0.621	kilometer, km (10^3 m)	mile, mi	1.609
1.094	meter, m	yard, yd	0.914
3.28	meter, m	foot, ft	0.304
1.0	micrometer, µm (10^{-6} m)	micron, µ	1.0
3.94×10^{-2}	millimeter, mm (10^{-3} m)	inch, in	25.4
10	nanometer, nm (10^{-9} m)	Angstrom, Å	0.1
Area			
2.47	hectare, ha	acre	0.405
247	square kilometer, km² (10^3 m)²	acre	4.05×10^{-3}
0.386	square kilometer, km² (10^3 m)²	square mile, mi²	2.590
2.47×10^{-4}	square meter, m²	acre	4.05×10^3
10.76	square meter, m²	square foot, ft²	9.29×10^{-2}
1.55×10^{-3}	square millimeter, mm² (10^{-3} m)²	square inch, in²	645
Volume			
9.73×10^{-3}	cubic meter, m³	acre-inch	102.8
35.3	cubic meter, m³	cubic foot, ft³	2.83×10^{-2}
6.10×10^4	cubic meter, m³	cubic inch, in³	1.64×10^{-5}
2.84×10^{-2}	liter, L (10^{-3} m³)	bushel, bu	35.24
1.057	liter, L (10^{-3} m³)	quart (liquid), qt	0.946
3.53×10^{-2}	liter, L (10^{-3} m³)	cubic foot, ft³	28.3
0.265	liter, L (10^{-3} m³)	gallon	3.78
33.78	liter, L (10^{-3} m³)	ounce (fluid), oz	2.96×10^{-2}
2.11	liter, L (10^{-3} m³)	pint (fluid), pt	0.473

CONVERSION FACTORS FOR SI AND NON-SI UNITS

Mass

To convert Column 1 into Column 2, multiply by	Column 1 SI Unit	Column 2 non-SI Unit	To convert Column 2 into Column 1, multiply by
2.20×10^{-3}	gram, g (10^{-3} kg)	pound, lb	454
3.52×10^{-2}	gram, g (10^{-3} kg)	ounce (avdp), oz	28.4
2.205	kilogram, kg	pound, lb	0.454
0.01	kilogram, kg	quintal (metric), q	100
1.10×10^{-3}	kilogram, kg	ton (2000 lb), ton	907
1.102	megagram, Mg (tonne)	ton (U.S.), ton	0.907
1.102	tonne, t	ton (U.S.), ton	0.907

Yield and Rate

0.893	kilogram per hectare, kg ha^{-1}	pound per acre, lb acre^{-1}	1.12
7.77×10^{-2}	kilogram per cubic meter, kg m^{-3}	pound per bushel, lb bu^{-1}	12.87
1.49×10^{-2}	kilogram per hectare, kg ha^{-1}	bushel per acre, 60 lb	67.19
1.59×10^{-2}	kilogram per hectare, kg ha^{-1}	bushel per acre, 56 lb	62.71
1.86×10^{-2}	kilogram per hectare, kg ha^{-1}	bushel per acre, 48 lb	53.75
0.107	liter per hectare, L ha^{-1}	gallon per acre	9.35
893	tonne per hectare, t ha^{-1}	pound per acre, lb acre^{-1}	1.12×10^{-3}
893	megagram per hectare, Mg ha^{-1}	pound per acre, lb acre^{-1}	1.12×10^{-3}
0.446	megagram per hectare, Mg ha^{-1}	ton (2000 lb) per acre, ton acre^{-1}	2.24
2.24	meter per second, m s^{-1}	mile per hour	0.447

Specific Surface

10	square meter per kilogram, m^2 kg^{-1}	square centimeter per gram, cm^2 g^{-1}	0.1
1000	square meter per kilogram, m^2 kg^{-1}	square millimeter per gram, mm^2 g^{-1}	0.001

Pressure

9.90	megapascal, MPa (10^6 Pa)	atmosphere	0.101
10	megapascal, MPa (10^6 Pa)	bar	0.1
1.00	megagram per cubic meter, Mg m^{-3}	gram per cubic centimeter, g cm^{-3}	1.00
2.09×10^{-2}	pascal, Pa	pound per square foot, lb ft^{-2}	47.9
1.45×10^{-4}	pascal, Pa	pound per square inch, lb in^{-2}	6.90×10^3

(continued on next page)

Conversion Factors for SI and non-SI Units

To convert Column 1 into Column 2, multiply by	Column 1 SI Unit	Column 2 non-SI Units	To convert Column 2 into Column 1, multiply by
Temperature			
1.00 (K − 273)	kelvin, K	Celsius, °C	1.00 (°C + 273)
(9/5 °C) + 32	Celsius, °C	Fahrenheit, °F	5/9 (°F − 32)
Energy, Work, Quantity of Heat			
9.52×10^{-4}	joule, J	British thermal unit, Btu	1.05×10^{3}
0.239	joule, J	calorie, cal	4.19
10^{7}	joule, J	erg	10^{-7}
0.735	joule, J	foot-pound	1.36
2.387×10^{-5}	joule per square meter, J m^{-2}	calorie per square centimeter (langley)	4.19×10^{4}
10^{5}	newton, N	dyne	10^{-5}
1.43×10^{-3}	watt per square meter, W m^{-2}	calorie per square centimeter minute (irradiance), cal cm^{-2} min^{-1}	698
Transpiration and Photosynthesis			
3.60×10^{-2}	milligram per square meter second, mg m^{-2} s^{-1}	gram per square decimeter hour, g dm^{-2} h^{-1}	27.8
5.56×10^{-3}	milligram (H$_2$O) per square meter second, mg m^{-2} s^{-1}	micromole (H$_2$O) per square centimeter second, µmol cm^{-2} s^{-1}	180
10^{-4}	milligram per square meter second, mg m^{-2} s^{-1}	milligram per square centimeter second, mg cm^{-2} s^{-1}	10^{4}
35.97	milligram per square meter second, mg m^{-2} s^{-1}	milligram per square decimeter hour, mg dm^{-2} h^{-1}	2.78×10^{-2}
Plane Angle			
57.3	radian, rad	degrees (angle), °	1.75×10^{-2}

Electrical Conductivity, Electricity, and Magnetism

10	siemen per meter, S m^{-1}	millimho per centimeter, mmho cm^{-1}	0.1
10^4	tesla, T	gauss, G	10^{-4}

Water Measurement

9.73 × 10^{-3}	cubic meter, m^3	acre-inch, acre-in	102.8
9.81 × 10^{-3}	cubic meter per hour, m^3 h^{-1}	cubic foot per second, ft^3 s^{-1}	101.9
4.40	cubic meter per hour, m^3 h^{-1}	U.S. gallon per minute, gal min^{-1}	0.227
8.11	hectare meter, ha m	acre-foot, acre-ft	0.123
97.28	hectare meter, ha m	acre-inch, acre-in	1.03 × 10^{-2}
8.1 × 10^{-2}	hectare centimeter, ha cm	acre-foot, acre-ft	12.33

Concentrations

1	centimole per kilogram, cmol kg^{-1}	milliequivalent per 100 grams, meq 100 g^{-1}	1
0.1	gram per kilogram, g kg^{-1}	percent, %	10
1	milligram per kilogram, mg kg^{-1}	parts per million, ppm	1

Radioactivity

2.7 × 10^{-11}	becquerel, Bq	curie, Ci	3.7 × 10^{10}
2.7 × 10^{-2}	becquerel per kilogram, Bq kg^{-1}	picocurie per gram, pCi g^{-1}	37
100	gray, Gy (absorbed dose)	rad, rd	0.01
100	sievert, Sv (equivalent dose)	rem (roentgen equivalent man)	0.01

Plant Nutrient Conversion

	Elemental	*Oxide*	
2.29	P	P$_2$O$_5$	0.437
1.20	K	K$_2$O	0.830
1.39	Ca	CaO	0.715
1.66	Mg	MgO	0.602

Introduction

Martin A. Massengale
Center for Grassland Studies
University of Nebraska
Lincoln, Nebraska

The purpose of this symposium was to assess our knowledge on several topics relating to warm-season grasses and to identify areas that need additional study. I commend the members of the symposium planning committee for assembling an excellent group of presenters that helped achieve that goal.

Before the early settlers arrived, native warm-season grasses dominated the tallgrass prairie region of North America, including the Midwest and Great Plains states. Today, most of these prairie areas have been converted to cropland or pastures dominated by cool-season grasses. The warm-season grasses are found in rangelands, unbroken prairie areas, or on lands under government agricultural programs such as the Soil Bank or Conservation Reserve. During the past two decades, however, warm-season grasses have become increasingly important as pasture grasses because of their ability to be productive during the hot summer months when cool-season grasses are relatively unproductive. Also, they are generally more drought tolerant.

There are several genera and species of warm-season grasses, but three grasses have dominated the North American tall grass prairie. They are big bluestem (*Andropogon gerardii* Vitman), indiangrass [*Sorghastrum nutans* (L.) Nash], and switchgrass (*Panicum virgatum* L.). These grasses have been used for more than a half century as pasture, range, and forage plants. Although these three grasses belong to different genera and differ in many ways, they do have a number of similarities such as end uses, management practices, and areas of adaptation. The distribution of warm-season grasses is affected greatly by temperature and moisture.

The western Corn Belt and Great Plains states are large livestock producing regions. Forages comprise a major component of this important enterprise and determine to a significant degree its profitability. One of the major problems confronting the livestock producer in these regions is the uneven distribution of forage production throughout the grazing season. Enough different grasses exist today that grazing systems can be developed to produce an abundance of high quality forage for the duration of the grazing season.

Cool-season grasses make their maximum growth and are heavy producers of forage in the spring and early summer. These plants usually become semi-dormant during hot weather, but growth is normally resumed during the cooler months of fall. In contrast, warm-season grasses start growth later in the spring, but grow

Copyright © 2000. Crop Science Society of America and American Society of Agronomy, 677 S. Segoe Rd., Madison, WI 53711, USA. *Native Warm-Season Grasses: Research Trends and Issues.* CSSA Special Publication no. 30.

throughout the summer and into early fall. Studies have shown that warm-season grasses produce a higher average daily gain than cool-season grasses during the summer months because of their succulent growth and better cattle nutrition. Some combination of these two groups of grasses is best for season-long grazing.

Warm-season grasses are well adapted to grazing, and with good grazing management it is no more harmful than no grazing. A sustained high level of productivity of these plants depends primarily upon management practices. When undergrazed, the forage is not used efficiently, and when overgrazed, the plants begin to lose their vigor, decline and die. The season of grazing also is highly important. Since these plants do not begin growing until late in the season, early removal of the foliage causes them to use their stored food reserves to produce more new foliage. If the crown and root reserves are repeatedly used to produce new growth early in the season, the plant will be weakened. If grazing is deferred until the plants are 8 to 10 inches tall, it should have no negative effect.

Despite the many potential uses and strengths of native warm-season grasses, they do have their problems. Research is needed to help solve them. Seedling establishment is a challenge. Warm-season grasses are known to be difficult to establish. Part of this problem can be attributed to seed dormancy, light, chaffy seed, and the lack of good seeding equipment. The establishment of warm-season grasses has improved in recent years because of improved seed quality, better seeding equipment, and improved herbicides for weed control.

Nutritive value of native warm-season grasses is often lower than that for cool-season grasses at comparable stages of maturity. Higher fiber and lower crude protein concentrations as well as lower leaf-to-stem ratios are involved. Plant breeding programs have emphasized both quality and yield in these grasses. Conventional methods of determining forage quality such as the percentage of digestibility and fiber and protein content are important to plant breeders who are working to improve these traits.

Generally, warm-season grasses and forage legumes have not been mixed because the forage legumes initiate growth earlier than the grasses, and this makes stands difficult to manage. When legumes and grasses are combined, however, the yield and protein content of the forage are increased.

Most warm-season native grasses do not tolerate severe and frequent defoliation as well as cool-season grasses. Therefore, invasion of weedy species is fairly common in these grass stands if they are mismanaged. For native warm-season grasses to achieve their productive potential, we must learn their unique physiological needs and develop cultural practices involving weed control, soil nutrient management, and cutting and grazing methods to capitalize on their strengths.

This publication examines topics such as seed dormancy, seedling establishment and developmental morphology, fertilization, cutting, grazing and weed management, breeding grasses for increased productivity, digestibility and protein content, establishing legumes in stands of warm-season grasses, and seed production in eastern gamagrass. Animal scientists and livestock producers will be particularly interested in the chapters on Escape and Rumen Digestible Protein Fractions along with Fiber Composition and Digestibility.

I commend this publication to you as an excellent source of information on several aspects of warm-season grass production and for your reading pleasure.

1 Escape and Rumen Degradable Protein Fractions in Warm-Season Grasses

Daren D. Redfearn

USDA Corn Insects and Crop Genetics Unit
Ames, Iowa

Karla Jenkins

University of Tennessee
Knoxville, Tennessee

Crude protein (CP) is routinely used to quantify protein concentration of forages. The CP concentration varies with forage species, plant part, plant maturity, and management practices. Protein degradation can be extensive in the rumen and often results in inefficient use of forage protein. This occurs because ruminal microorganisms, which can both synthesize and degrade protein, degrade a larger amount than is synthesized; therefore, much forage protein is wasted.

The dynamic nature of plant–animal interactions and the inability to easily or accurately analyze the grazed diet precludes timely forage protein analysis. Generally, percentage of CP decreases with maturity while animal selection increases to compensate for this decline. Supplementation requirements on pasture generally address protein deficiencies and will become more sophisticated as plant protein composition is further defined by species and maturity. Animal performance on warm-season (C_4) pastures with marginal CP levels, however, is similar to performance on cool-season (C_3) pastures. Although CP values may appear adequate, rumen degradable or escape protein may be deficient during certain periods of the growing season.

Crude protein may be insufficient to balance animal rations, particularly for high-producing ruminants. When supplemental protein is fed, the degradability of the supplemental protein (rumen degradable or rumen escape) with respect to the forage must be considered. Animals that graze cool-season pastures usually benefit if supplemented with rumen escape protein, while animals grazing warm-season pastures generally do not benefit when supplemented with additional escape protein. It is not unrealistic to think that animals consuming warm-season grasses may be deficient in rumen degradable protein and require additional rumen degradable protein for maximum performance.

Copyright © 2000. Crop Science Society of America and American Society of Agronomy, 677 S. Segoe Rd., Madison, WI 53711, USA. *Native Warm-Season Grasses: Research Trends and Issues.* CSSA Special Publication no. 30.

RUMINANT NITROGEN USAGE

Use of dietary N differs between those animals that rely on a hydrolytic digestive process (nonruminant) and those that depend on a fermentive-type digestion (ruminant). Dietary protein ingested by ruminants is degraded in the rumen to various extents by rumen microorganisms (Fig. 1–1). Factors that influence differences in amino acid degradation are the nature and solubility of the protein, rate of passage of the digesta, and level of intake (Tamminga, 1979).

Ruminants derive their amino acid supply from microbial protein synthesized in the rumen, undegraded proteins, and amino acids that escape ruminal degradation. Protein that is protected from ruminal microbial degradation allows more amino acids to reach the small intestine, thus increasing animal performance (Chalupa, 1975). This occurs because ruminal protein degradation results in inefficient use of forage protein and suppressed animal performance. Escape protein is required to supply required metabolizable protein in excess of microbial protein production.

Metabolizable Protein

Protein needs of ruminants are met by microbial protein synthesis and by dietary forage protein that escapes ruminal degradation and is absorbed from the small intestine (Burroughs et al., 1974; National Research Council, 1985). Some forages provide highly degradable protein, while others provide more slowly degraded proteins. Thus, metabolizable protein is be a better expression of protein value than CP in ruminant diets (Fig. 1–2).

Crude protein and digestible protein may provide adequate estimates of the protein requirement for ruminants in early pregnancy, maintenance, or growing at a slow rate (low-producing), but these same estimates are most likely inappropriate for animals in the third trimester of pregnancy, early lactation, or growing at a fast rate (high-producing). If amino acid requirements exceed that provided by microbial protein synthesis, a portion of the dietary intake protein must escape rumen degradation and be available post-ruminally.

Microbial Protein

In addition to N, energy must be present for adequate microbial protein synthesis. Nitrogen sources must be balanced with energy for efficient microbial protein synthesis (Burroughs et al., 1975), which is a major component of metabolizable protein. Microbial protein reaching the small intestine is 20 to 50% of the DM and actually can provide more protein than the ingested forage contained when the forage is low in protein. Accumulation of ruminal ammonia occurs when forage protein is fermented and deaminated. A portion of the ammonia is absorbed across the rumen wall and into the bloodstream and subsequently detoxified into urea by the liver; however, a portion of the urea is recycled to the rumen via saliva and reabsorbed through the rumen wall. Substantial amounts are excreted in the urine (Minson, 1990), especially in cattle consuming forages with great amounts of rumen degradable protein (Fig. 1–1). If the ingested protein is poor quality, microbial protein synthesis may enhance protein quality of the diet (Owens & Zinn, 1988).

ESCAPE AND RUMEN DEGRADABLE PROTEIN

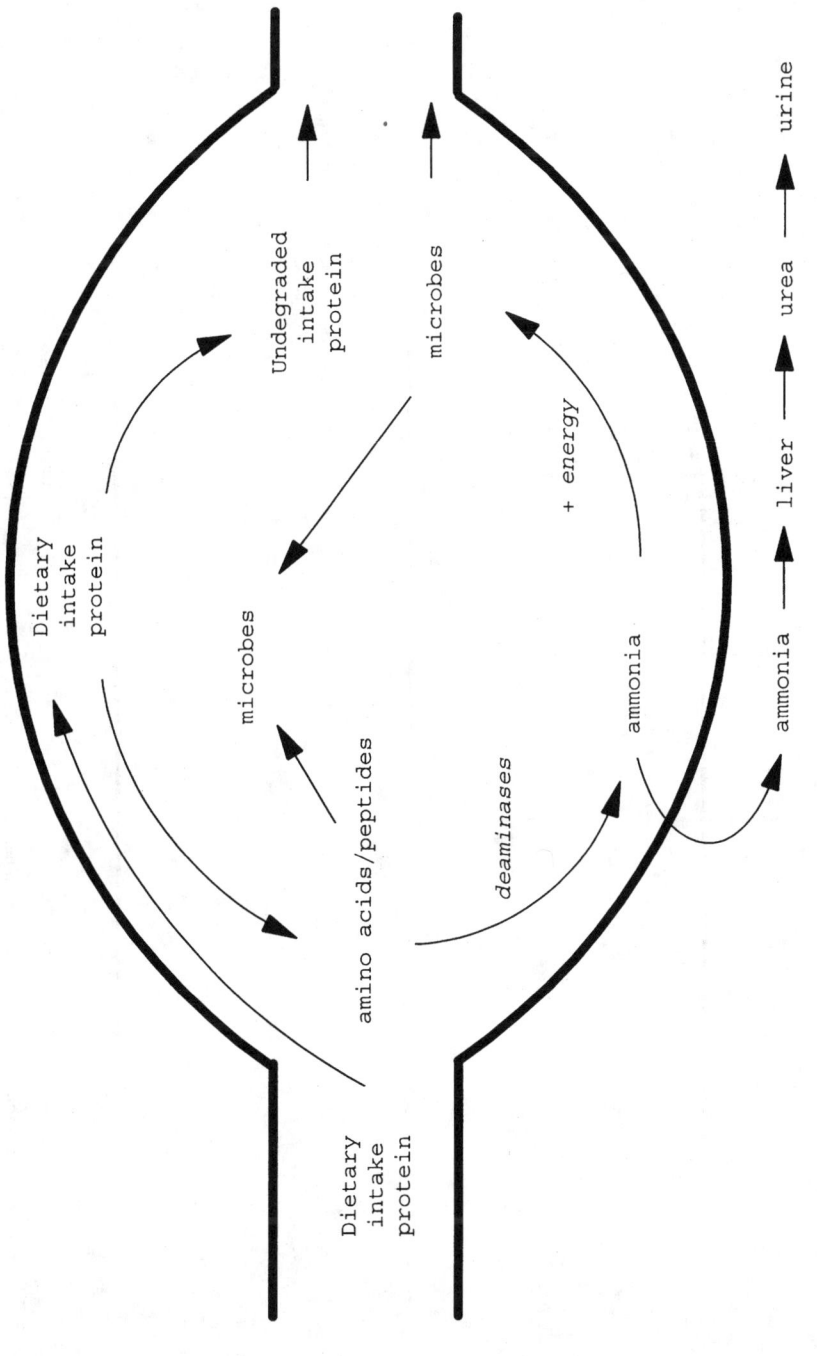

Fig. 1–1. Diagram of the fate of dietary intake protein and ruminal microbial protein synthesis.

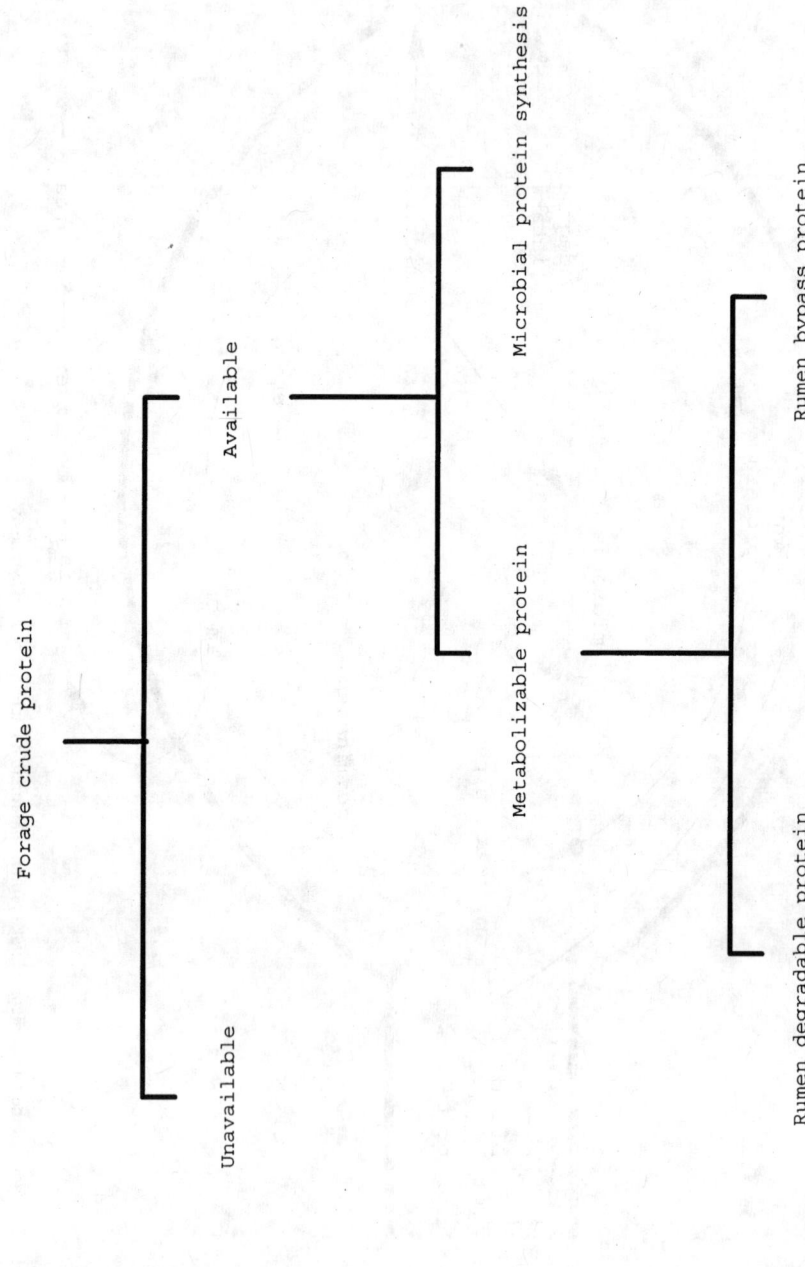

Fig. 1–2. Conceptualization of forage crude protein partitioning.

Microbial needs must be met so that microorganisms can provide adequate fiber digestion, volatile fatty acid production, and production of C skeletons from amino acids for microbial protein synthesis. Microbial N needs are met by ammonia, amino acids, and peptides. Ammonia is the most important intermediate in N usage of ruminant microorganisms (Annison & Lewis, 1959; Bryant & Robinson, 1963; Stevenson, 1978). Duncan et al. (1953) established that the amino acid profile of ruminal contents is similar to that of ruminants fed protein diets. Thus, ruminants supply a relatively constant quality of protein to the small intestine regardless of the dietary protein quality (Steinhour & Clark, 1980). In maintenance production systems with gestating beef cows (*Bos taurus*) grazing warm-season grasses, microbial protein flow and basal dietary escape protein are usually adequate to meet the animal's protein requirement.

FORAGE PROTEIN COMPOSITION

Crude protein has long been used to quantify the protein content of forages. It is estimated as Kjeldahl N × 6.25 and is based on the assumption that the average N concentration of all proteins is 16%. Crude protein is a good indicator of forage quality within a species, but is not acceptable for comparisons across species. Therefore, measurements of CP usually are inadequate to characterize and classify total protein for nutritional purposes of grazing ruminant animals (Mangan, 1982).

The CP of forages is comprised of true protein and nonprotein nitrogen (NPN). About 75% of the N in leaves is true protein. Plant true proteins function as enzymes and thus, are important to the ability of the plant to carry on photosynthesis, respiration, and growth. True protein in fresh forages has been classified into three major classes: Fraction I leaf protein, Fraction II proteins, and chloroplastic membrane proteins (Mangan, 1982) based on their apparent solubility in water. Fraction I protein is water-soluble, whereas Fraction II contains both water-soluble and water-insoluble protein fractions. The chloroplastic membrane proteins are water-insoluble, partly because of the hydrophobic bilipid layer located in the chloroplast membrane.

Leaf proteins of cool-season species are well-characterized. Fraction I protein is a homogenous chloroplastic protein made up of the macromolecule ribulose-1,5-bisphosphate carboxylase-oxygenase (RuBPcase). Huffaker (1982) estimated that Fraction I protein accounted for 40 to 80% of the total soluble protein in C_3 species found in the chloroplasts of the mesophyll cells. Fraction II proteins constitute about 25% of leaf true proteins. Fraction II proteins that have been characterized and isolated in large quantities include ferredoxin, plastocyanin and the cytochromes of chloroplasts and tubulin, actin, ATP synthetase, carbonic anhydrase, and protein elongation factors from the cytoplasm. Proteins associated with the chloroplast lamellar (thylakoid) membranes are insoluble in water and constitute about 40% of the chloroplast protein (Mangan, 1982). Thylakoid membrane protein is further divided into Chlorophyll Complex I and Chlorophyll Complex II (Mangan, 1982). Free amino acid N averaged approximately 10% of protein in forages (Ferguson & Terry, 1954). The NPN fraction comprises approximately 25% of the total N in leaf blades, although the NPN component may range from 15 to

50% of the total N on a whole-plant basis. This fraction predominately consists of DNA, RNA, mitochondrial proteins (Lyttleton, 1973), a hydroxyproline rich cell-wall glycoprotein (extensin) firmly bound to cellulose (Lamport & Northcote, 1960), and free amino acids, as well as nitrates, ammonia, and amides. The primary function of the NPN fraction is to serve as intermediates of protein synthesis, agents of translocation, and as products of inorganic N assimilation (Mangan, 1982).

While protein fractions of cool-season grasses have been well-characterized, much less is known about the protein fractions in warm-season grasses. Fraction I protein makes up only 4 to 8% of soluble leaf protein in C_4 species (Brown, 1978) and is restricted to the thick-walled bundle sheath cells; however, about 50% of the total leaf protein is located in the bundle sheath cells. Mesophyll cells of warm-season species contain no Fraction I protein in the chloroplasts. Phosphoenol pyruvate carboxylase replaces RuBPcase in the mesophyll of warm-season plants (Brady, 1976) and may comprise 10 to 15% of the true protein in the leaves (Uedan & Sugiyama, 1976). Since minimal information exists regarding nitrogenous fractions in warm-season grasses, no additional conclusions can be drawn concerning protein fractions.

Generally, whole-plant CP concentration is inversely related to stage of plant maturity. Ranges in CP concentration for most cool-season grasses are 120 to 180, 90 to 120, and 60 to 90 g kg^{-1} for vegetative, elongating, and reproductive maturities, respectively. Ranges in CP concentration for warm-season grasses at a similar stage of maturity are usually lower than for cool-season grasses, although some overlap may occur. Typical ranges for CP concentration of warm-season grasses at vegetative, elongating, and reproductive sward maturities are 100 to 150, 70 to 100, and 40 to 70 g kg^{-1}, respectively.

DEGRADATION OF FORAGE PROTEINS

Simplistically, forage protein fractions can be divided into two fractions: unavailable and available (Fig. 1–2). The unavailable fraction can be estimated as acid detergent insoluble N (ADIN). This fraction is assumed to be completely indigestible (Van Soest, 1982). Acid detergent insoluble N is interpreted to be N that is neither rumen degradable nor digested in the small intestine. Overall, it is a good measure, but not perfect. The available fraction is comprised of metabolizable forage protein and synthesized microbial protein. The metabolizable fraction derived from forage protein is comprised of all soluble forage N and neutral detergent fiber (NDF) N. For cell soluble forage N, we typically assume that it is completely rumen degradable, but this is most likely inaccurate. Additionally, it is also difficult to speculate about N type, such as ammonia, urea, amino acids, or peptides.

New systems for calculating protein requirements for grazing ruminants are necessary. Historically, measuring extent of protein degradation has been used only to evaluate forages by ranking treatments consisting of species, cultivars, or differences in plant maturity. Rate of forage intake and rate of passage, however, must be incorporated into any system to completely characterize rumen degradable and escape protein estimates. A more recent model for predicting metabolizable protein is the Cornell Net Carbohydrate and Protein System (CNCPS) (Sniffen et al.,

1992). The CNCPS assumes that metabolizable protein is variable for all forages based on protein composition, ruminal protein digestion rates, passage rates, microbial protein yield and composition, and postruminal digestibilities of dietary and microbial protein. The CNCPS assumes five protein fractions (A, B1, B2, B3, and C), where Fraction A is assumed to be NPN and Fraction C is bound true protein. Fractions B1, B2, and B3 are assumed to be true protein fractions degraded in the rumen in a rapid, intermediate, and slow manner, respectively.

The CNCPS does not use total digestible nutrient (TDN) values as a measure of energy that is available for microbial protein synthesis. The CNCPS is based on the concept that microorganisms do not use protein, lipids, or ash as a source of energy (Nocek & Russell, 1988). Use of TDN values as estimates of available energy assumes that all of the digestible energy is available to the microorganisms and will overestimate microbial protein synthesis. The CNCPS is based on carbohydrate digestion by two groups of microorganisms. Structural carbohydrate fermenting bacteria degrade cellulose and hemicellulose and are assumed to use ammonia as their N source. Nonstructural carbohydrate bacteria degrade starchs, pectins, and sugars and are assumed to use either ammonia or amino acids as their N source (Russell et al., 1992). Although it is oftentimes comforting to allow a mechanistic model to predict values in complex biological systems, the predicted values should be evaluated for use only as potential recommendations.

For cool-season grasses, Muscato et al. (1983) reported that mature timothy (*Phleum pratense* L.) hay had 7.8 g N kg^{-1} NDF and 2.5 g N kg^{-1} ADF; however, no digestibility data were reported. Hogan and Lindsay (1979) reported that immature ryegrass (*Lolium perenne* L.) contained 30.1 g N kg^{-1} cell wall and accounted for approximately one-fourth of the total forage N. In immature ryegrass, they also found 81% of the organic matter and 89% of the cell-wall N to be rumen degradable when fed to sheep (*Ovis aries*). They concluded that negligible amounts of amino acids from forage cell-wall protein were digested and absorbed in the small intestine.

Sanderson and Wedin (1989a) reported that cell walls in leaf blades of timothy and smooth bromegrass (*Bromus inermis* Leyss.) contained 8.6 and 6.4 g N kg^{-1} NDF, respectively, and 3.2 and 2.4 g N kg^{-1} NDF, respectively, for stems. Weiss et al. (1986) found fresh alfalfa (*Medicago sativa* L.) contained 12 g N kg^{-1} NDF and 6 g N kg^{-1} ADF. Apparent digestibilities were 39 and 31% for NDFN and ADIN, respectively. There are virtually no data on the concentration and digestibility of N in cell-wall fractions of warm-season forage grasses. For warm-season grasses, Amos et al. (1984) found that 'Coastal' bermudagrass [*Cynodon dactylon* (L.) Pers.] contained 10 g NDFN kg^{-1} DM and 1 g ADIN kg^{-1} DM, whereas alfalfa contained 4 g NDFN kg^{-1} DM and 2 g ADIN kg^{-1} DM. Rumen degradability of NDFN was 84 and 57% for bermudagrass and alfalfa, respectively; however, digestion coefficients for NDFN were not corrected for microbial protein or ADIN. If these values had been corrected, the apparent digestion coefficients would have been higher.

For cool-season grasses, >80% of the total N is in cell solubles (Sanderson & Wedin, 1989b). Cell wall N concentrations are generally lower for cool-season grasses than legumes, but cell wall N accounts for a greater proportion of the total N in grass leaves than legume leaves. This is because grass leaves have higher cell wall concentrations than legume leaves (Buxton et al., 1996). To date, no information is available for similar estimates in warm-season grasses; however, it would be expected that a similar relationship would exist for warm-season grasses since they typically have greater cell wall concentrations than cool-season grasses.

Hunter and Siebert (1985) predicted that 20 to 35% of the initial cell solubles of the warm-season grasses, speargrass [*Heteropogon contortus* (L.) Beauv. ex Roem. & Schult.] and pangolagrass (*Paspalum dilatatum* Poir.), would remain intact following rumen digestion. This may be due in part to protection of the cell solubles by the well-developed parenchyma bundle sheath cells that tended to be more slowly digested than the mesophyll and phloem in C_4 species (Akin & Burdick, 1975).

Mitchellgrass (*Astrebla pectinata* F. Muell.), a C_4 species, had 72 to 82% ruminal protein degradability (McMeniman et al., 1986), but these values were unadjusted for endogenous N flow so the estimates may be as much as eight percentage points too low. Beever et al. (1986) found that degradability of ryegrass protein ranged from 64 to 82%. Ruminal N degradation of various ryegrasses produced values that ranged from 50 to 70% (Beever et al., 1974; Walker et al., 1975). Mullahey et al. (1992) estimated rumen escape protein in leaf and stem fractions of switchgrass (*Panicum virgatum* L.) and smooth bromegrass. Escape protein concentration expressed as a percentage of the total CP in topgrowth switchgrass was higher (51%) than smooth bromegrass (19%). Concentrations of escape protein were greater in leaves than stems.

While energy digestibility is similar for fresh and dried forages, N degradability may be markedly different. Beever et al. (1971) found apparent N degradability to be 75% for fresh ryegrass and 71% for the same ryegrass hay. Beever et al. (1974) found that ryegrass frozen before feeding showed a 15% increase in the amount of amino acids entering the small intestine and a 26% increase in microbial protein synthesis over fresh ryegrass. Conversely, dried ryegrass showed an increase of 51 and 57% for amino acids and microbial protein synthesis, respectively, compared with fresh ryegrass. While the nondegraded protein fraction may be increased when forage is preserved as hay, protein in fresh forage typically is highly rumen degradable. When nonammonia N is abundant, however, its apparent loss across the rumen wall may be as high as 30% (Beever & Siddons, 1986). Due to limits in available energy and digestibility of microbial N, most of the ingested N is not available to the animal (Anderson et al., 1988). Therefore, while fresh forage may be high in protein quantity, escape protein may be needed to meet metabolizable protein requirements.

ESCAPE PROTEIN

Rumen escape protein is the nondegraded or protected dietary proteins, amino acids, and peptides. Although addition of dietary escape protein may enhance production and/or reduce production costs, it cannot be used to completely replace rumen degradable protein. As mentioned earlier, microbial protein synthesis requires ammonia, amino acids, and peptides. These precursors to microbial protein production can be limiting in forages containing proteins slowly degraded in the rumen or low in CP. A minimum rumen ammonia or CP level is needed for basal microbial protein synthesis. Therefore, it is essential to supply rumen degradable proteins in combination with high escape protein forages when rumen degradable protein of warm-season forage diets may be deficient.

Forage escape protein values are usually expressed either on a dry matter (DM) or CP basis. Escape protein expressed as a percentage of DM is the actual amount of escape protein. Alternatively, escape protein expressed as a percentage of CP only describes the percentage of CP that has escape protein value and contributes nothing to determining protein deficiencies in ruminant diets. As plant maturity increases with the concomitant decrease in CP, the amount of escape protein remains relatively constant; however, the dilution of the overall CP results in a dramatic inflation of escape protein value. This illustrates the point that just because a greater percentage of escape protein is present more protein may not necessarily be available because of an inherently low CP concentration. A lack of understanding between the two may result in misinterpretation of data and nutritional significance.

Anderson et al. (1988) observed a 0.15 kg d^{-1} increased response to escape protein supplied as corn gluten meal (55% escape protein) and bloodmeal (82% escape protein) to steers grazing smooth bromegrass. Blasi et al. (1991) showed an increase in milk production of lactating beef cows grazing smooth bromegrass when individually supplemented with an escape protein mixture of bloodmeal and corn gluten meal and as a result, increased calf weights. Lactating beef cows individually supplemented with the same escape protein mixture showed no response while grazing a big bluestem (*Andropogon gerardii* Vitman) pasture. Anderson et al. (1988) found that while rumen degradable protein was more than adequate in smooth bromegrass, there was still a need for supplemental escape protein.

This response to escape protein by lactating beef cows grazing smooth bromegrass pastures suggests that metabolizable protein may be limiting animal performance. Anderson et al. (1988) calculated that only 9.5 to 13.5% of the protein in smooth bromegrass would escape rumen digestion at a digestion rate of 13.6% h^{-1}. Approximately 50% of the forage protein from a mixed stand of big bluestem, switchgrass, and indiangrass [*Sorghastrum nutans* (L.) Nash] escaped rumen degradation (Hafley et al., 1993). Similar results were reported by others for escape protein concentration of warm- and cool-season grasses (Table 1–1).

The need for dietary escape protein is secondary to rumen microbial N needs and may be most critical for ruminants grazing warm-season grasses. With growing calves, grazing yearlings, and lactating beef cows, however, escape protein is essential for maximum production and may be more economical than feeding large amounts of rapidly degraded protein supplements. The amount of supplemental dietary protein needed depends on animal production status and forage metabolizable protein concentration. For example, while growing calves grazing cool-season grasses typically respond to escape protein supplementation, Hollingsworth et al. (1993) observed no response to escape protein for growing calves grazing sudangrass [*Sorghum sudanese* (Piper) Staph.], an annual warm-season grass. Laboratory analyses indicated the forage contained adequate metabolizable protein for growing calves. Similarly, Hafley et al. (1993) observed that gains of yearling steers grazing a mixed stand of big bluestem, switchgrass, and indiangrass supplemented with a feathermeal (65% escape protein) and bloodmeal mixture were similar to non-

Table 1–1. Crude protein and escape protein concentrations of warm- and cool-season grasses.

Species	Crude protein	Escape protein – Crude protein basis	Escape protein – Dry matter basis
		g kg^{-1}	
Warm-season			
Switchgrass[†]	75	541	--
Switchgrass[‡]	77	509	32
Switchgrass[§]	111	234	25
Big bluestem[†]	67	708	--
Big bluestem[§]	104	262	28
Big bluestem[¶]	128	247	28
Indiangrass[†]	68	591	--
Mixed species[#]	71	567	38
Cool-season			
Smooth bromegrass[‡]	111	205	22
Smooth bromegrass[§]	133	150	20
Smooth bromegrass[¶]	210	107	21
Intermediate wheatgrass[§]	121	118	14

[†] Cuomo and Anderson, 1996. No adjustments for ADIN or microbial protein.
[‡] Mullahey et al., 1992. Adjustment for ADIN but not for microbial protein.
[§] Mitchell et al., 1997. Adjustments for both ADIN and microbial protein.
[¶] Blasi et al., 1991. Adjustment for ADIN but not for microbial protein.
[#] Hafley et al., 1994. Adjustment for ADIN but not for microbial protein.

supplemented animals. Rumen degradable protein, however, increased gains 0.09 kg d^{-1} over nonsupplemented animals, whereas rumen degradable protein in combination with rumen escape protein increased gains up to 0.13 kg d^{-1} compared with nonsupplemented animals.

PROTEIN DEGRADATION TECHNIQUES

The information reported in this chapter is not meant to an encompassing review and critique of all research pertaining to protein degradation techniques. Broderick (1994) has previously discussed the advantages and disadvantages using most of the techniques summarized in this article. Protein degradation can be estimated by in vitro, in situ, or in vivo methods. The majority of protein degradation techniques were not developed to estimate forage protein degradability, but rather for protein supplements or individual feed sources. Numerous attempts, however, have been made to modify procedures for estimating forage protein degradation. Newer, innovative techniques such as sodium dodecyl sulfate-polyacrylamide gel electrophoresis (SDS-PAGE) and immunochemical techniques are being used more frequently to characterize forage proteins.

In vitro

The most inexpensive and widely used method to estimate ruminal protein degradation is a modified version of the Tilley and Terry (1963) technique that correlates ammonia release with ruminal protein degradability, where greater ammonia concentrations indicate increased ruminal protein degradation. In vitro methods include ammonia release and enzymatic digestion (Britton et al., 1978; Poos-Floyd et al., 1985). Nitrogen solubility also can be used as a method of in vitro estimation of ruminal protein degradability. Ruminal protein degradation has been estimated by solubility using autoclaved rumen fluid, hot water, 0.02 M NaOH, bicarbonate-phosphate buffer, and 0.1 M NaCl (Poos-Floyd et al., 1985). Caution must be exercised when interpreting N solubility values. Nitrogen solubility is not synonymous with rumen degradability. Protein structure such as disulfide linkages also can affect rumen degradability. Although N solubility may be well-correlated with rumen degradability, using solubility values as a substitute for degradability values may be incorrect and misleading. Since most N solubility research is done with concentrated protein sources, these methodologies may not be suitable for determining escape protein in forages.

Proteases also have been examined as an in vitro method for estimating protein degradability. Poos-Floyd et al. (1985) determined that among neutral bacterial proteases (papain, bromelin, ficin), and a fungal protease, ficin and the neutral fungal protease were most highly correlated with in vivo degradation. Enzymatic protein degradation may yield different results when used with forages rather than with concentrated protein sources. Therefore, research using enzymatic procedures as a method for determining forage protein degradability is limited (Krishnamoorthy et al., 1982).

In situ

Polyester bags are popular have been widely used to predict protein degradability in the rumen. Mehrez and Orskov (1977) found polyester bags to be satisfactory as a guide for measuring nutrient disappearance in the rumen. The in situ procedure involves incubating forages in polyester bags in the rumen for different periods of time to establish protein degradation rates or escape protein estimates. This method requires rather inexpensive materials, but is somewhat labor intensive. Earlier in situ determinations used a 16-h incubation for forage escape protein analysis. Broderick (1993), however, suggested the use of two to three incubation times and an estimate of passage rate to accurately determine absolute escape protein values rather than relative estimates.

Following incubation, the bags must be thoroughly rinsed and dried with subsequent Kjeldahl analysis and corrections for microbial attachment. Although the method is fairly simple, incorrect procedures result in large variations in escape protein estimates. Wilkerson et al. (1995) showed the rinsing method to be a critical factor in reducing overall error. Microbial contamination may bias results such that the escape protein value is inflated.

Methods for estimating microbial protein contributions are difficult to development and have been somewhat unreliable. Early attempts at releasing attached bacteria were made by Minato et al. (1966). They found that rumen contents taken at 4 and 24 h after ingestion contained less protein after washing in a salt water solution than unwashed rumen contents. Later attempts at releasing attached bacteria with surfactants and salt solutions also were made by Akin (1980), Dehority and Grubb (1980), and Craig et al. (1987). Mechanical means for separating attached bacteria also have been tried without much success (Mackie et al., 1983; Merry &

McAllen, 1983). Markers have also been employed to estimate microbial attachment to forage particles after incubation. Enrichment or infusion with ^{15}N and cytosine have been used to estimate microbial contamination (Wanderley et al., 1993). The use of purines as a microbial marker has been described by Zinn and Owens (1986) and Aharoni and Tagari (1991).

In vivo

In vivo estimates are probably the most accurate, but also are most costly and time consuming. This method requires abomasal or duodenal cannulas, as well as reliable digesta flow markers. There are two ways to measure protein degradation in vivo. The first is using a regression method, in which the protein is added incrementally to a base ration of constant DM and fermentability, with one diet providing only enough N for microbial protein synthesis. Amino acid flows from the duodenum are regressed against amino acid intake. Thus, the slope of the regression equation represents the undegraded protein. The second method of estimating protein degradability is by difference. This technique is more direct in that dietary intake protein is known and the total duodenal protein flow is measured. Microbial and endogenous N sources are estimated at the duodenum so that undegraded dietary protein is calculated as the difference. Thus, undegraded protein = [duodenal protein − (microbial protein + endogenous protein)] (Stern & Satter, 1980).

Electrophoresis and Immunochemistry

Although electrophoretic and immunochemical techniques are not new, their use in protein evaluation for ruminants has been quite limited. SDS-PAGE separates protein fractions solely on the basis of their molecular weight. Jones and Mangan (1976) used SDS-PAGE to separate heat-treated Fraction I leaf protein from alfalfa. Fraction II leaf proteins have been separated into 6 to 9 major, but unidentified peaks (Mangan, 1982). Cherney et al. (1992) separated soluble protein fractions of alfalfa leaves and stems in a variety of solvents using SDS-PAGE.

Information on protein degradation using these innovative techniques has only recently taken precedence over the more conventional CP analysis. Redfearn et al. (1995) used SDS-PAGE to follow forage grass leaf protein degradation in switchgrass, big bluestem, and smooth bromegrass. Even though the identity of proteins were not established with certainty, they concluded that the increasing proportions of proteins escaping ruminal degradation suggested that some mechanism was present in the warm-season grasses that allowed certain proteins to be undegraded for longer periods compared with smooth bromegrass. Miller et al. (1996) followed the loss of RUBPcase from switchgrass and big bluestem using immunofluorescent localization. This technique uses antisera produced in response to a specific protein to detect for presence or absence of that protein.

While each of these two methods has advantages and disadvantages, the preferred method for forage protein analysis may be immunoblotting that combines electrophoretic and immunochemical techniques. The distinct advantage that this method offers is that site, rate, and extent of degradation for specific proteins could be determined. Although more labor intensive and costly than conventional CP analysis, combining these two techniques will undoubtedly improve our understanding of plant protein digestion and utilization for ruminant animals.

MANAGEMENT

Warm-season grasses typically have lower CP concentrations than cool-season grasses and legumes at comparable stages of maturity. Crude protein concentrations are usually greatest in the leaves and lowest in the stems. As plant maturity increases, CP concentration usually decreases. Warm-season pasture renovation using a combination of prescribed burning and fertilizer applications may substantially increase CP concentration and alter protein degradability of available forage. These changes in protein degradability are attributed more to the changes that occur during plant development that influence leaf to stem ratio than by chemically altering protein composition and quality.

Prescribed Burning

Mitchell et al. (1994) evaluated CP response in big bluestem to prescribed burning at delayed intervals (Table 1–2) and found that late-spring burning increased early-summer CP concentration by 15% compared with no burning. Late-spring burning also increased mid-summer CP by 8% compared with early-spring burning. Although CP concentration was increased by prescribed burning, differences in CP can be explained by the maturity differences of the burning treatments. Burning increased early season CP in switchgrass, big bluestem, and indiangrass, but rumen degradable protein was unaffected by prescribed burning (Cuomo & Anderson, 1996).

Even though prescribed burning had no direct influence on forage protein degradability, it did influence plant growth and development, which is related to protein degradability. Mullahey et al. (1992) found that whole-plant escape protein expressed as a percentage of total CP was greater for switchgrass (50.9%) than smooth bromegrass (20.5%). Whole-plant escape protein expressed on a DM basis, however, was greater for switchgrass (31.8 g kg^{-1} DM) than smooth bromegrass (22.3 g kg^{-1} DM) averaged across several growth stages. Escape protein expressed as a percentage of the total DM decreased in both species as plant maturity increased because of decreased leaf to stem ratio. This occurred in spite of escape protein percentage of smooth bromegrass leaves and stems having similar escape protein percentage, whereas switchgrass stems had greater escape protein than leaves.

Table 1–2. Crude protein concentration of big bluestem after spring burning of tallgrass prairie.

Spring burning	June	July	August
		g kg^{-1}	
None	101	70	52
Early	91	68	47
Mid	100	69	47
Late	119	74	45
LSD (0.05)		5	

Table 1–3. Crude protein concentration of big bluestem after fertilizing tallgrass prairie.

Year	Fertilizer	June	July	August
	kg N–P ha^{-1}		g kg^{-1}	
1989	0	85	62	50
	67–22	104	76	58
1990	0	100	66	41
	67–22	122	77	42
LSD (0.05)			4	

Nitrogen

Forage CP and protein fractions are greatly affected by soil N fertility levels. Nitrogen fertilizers increased the NPN content of cool-season grasses (Lexander et al., 1970) with true protein percentage in leaves decreasing due to an increase in free amino acids that consist predominately of asparagine, glutamine, aspartic acid, glutamic acid, alanine, and serine. Ammonia fertilization reduced true protein N to 60 to 65% and nitrate fertilization further reduced it to 42 to 46% (Mangan, 1982). The nitrates must be converted to ammonia and then to amino acids before assimilation into true protein (Salisbury and Ross, 1985). Fertilization of warm-season grass monocultures, including 'Trailblazer' switchgrass, 'Pawnee' big bluestem, and 'Nebraska 54' indiangrass increased CP concentration from 64 g kg^{-1} in nonfertilized plots to 76 g kg^{-1} in plots receiving 66 kg N ha^{-1} (Cuomo & Anderson, 1996). Approximately two-thirds of increased CP from fertilization was rumen degradable protein. This response was similar to the results of George and Farnham (1991) who reported that rumen degradable protein fraction of switchgrass increased when 120 kg N ha^{-1} was applied compared with nonfertilized switchgrass.

Mitchell et al. (1994) also evaluated CP responses of big bluestem to N fertilizer applications (Table 1–3). Although estimates of protein degradability were not determined, it is assumed that big bluestem protein fractions responded similarly to the degradability estimates reported by Cuomo and Anderson (1996).

SYNTHESIS

Several factors may contribute to the anomaly of similar animal performance for cattle grazing warm-season grasses compared with cattle grazing cool-season grasses. These may include, but not be limited to differences in ruminal microbial populations, chemical differences occurring as a normal developmental process, and the N use efficiency of ruminants consuming warm-season grasses. It is not unrealistic to think that at least two, if not all of these factors interact to confound our attempts to answer questions concerning rumen degradable and bypass protein fractions of warm-season grasses.

Rumen microbial populations of cattle grazing warm-season grasses may be vastly different from cattle grazing cool-season pastures. Major cellulolytic species (cell wall fermenters) in the rumen include *Bacteroides succinogenes*, *Ruminococcus flavefaciens*, *R. albus*, and *Butyrivibrio fibrisolvens*. Major proteolytic species (protein degraders) include *B. amylophilus*, *B. ruminicola*, *B. fibrisolvens*, and *Streptococcus bovis*. Since these microorganisms prefer a different substrate, different populations of each may occur in the rumen. Because the fiber in cool-season forages is more rapidly fermented and the majority of the forage protein is associated with this rapidly fermented substrate, all of the major cellulolytic and proteolytic microorganisms may be prevalent in the rumen environment. With warm-season forage species, fiber digestion proceeds more slowly. If the protein associated with the more slowly digested fiber is not liberated, the major proteolytic species may not be as numerous as in the rumen environment with cool-season grasses. Therefore, if these proteolytic species are not present or are present in significantly lower numbers, a portion of the forage protein may not be completely degraded in the rumen.

Although grasses do not contain tannins per se, flavonoid compounds (anthocyanins), which are tannin-like, do accumulate in leaves of warm-season grasses (Salisbury & Ross, 1985). Legumes that contain condensed tannins, such as large trefoil (*Lotus pedunculatus* Cav.) (Barry & Manley, 1984) and sainfoin (*Onobrychis viciifolia* Scop.) (Mangan et al., 1976), have shown reduced ruminal protein degradation. Barry and Manley (1984) observed an increase in nonammonia N digested post-ruminally by sheep fed large trefoil that contained high tannin concentrations compared to sheep fed large trefoil that contained low tannin concentrations or grazed ryegrass. Mangan et al. (1976) fed sainfoin to cattle and observed no release of soluble N or chlorophyll. They concluded that the protein formed complexes with the condensed tannins and the protein was not released by mastication.

The tannin-protein complex is insoluble at pH 3.5 to 7.0, but is soluble at both lower and greater pH. Rumen pH of forage-fed animals typically ranges from approximately 6.5 to 7.0. This complex escapes rumen degradation and is hydrolyzed in the abomasum (pH 2.5 to 3.5) and small intestine (pH 7.5 to 8.5). Thus, if anthocyanins or other flavonoid compounds behave similarly to tannins in the rumen, this may partially explain a portion of the perplexing results concerning protein nutrition of animals grazing warm-season grasses.

Lastly, efficiency of N utilization may be increased in cattle grazing warm-season pastures. Because protein fractions of cool-season grasses are more rapidly degraded in the rumen than warm-season grasses, differences in accumulation of ammonia exist (Hafley et al., 1993). Redfearn et al. (1995) found that rate of protein disappearance for switchgrass was slower than for big bluestem, but big bluestem and smooth bromegrass disappeared at similar rates and switchgrass disappeared more slowly than smooth bromegrass. They also identified three distinctive protein fractions that were slowly degraded in all three species; however, these protein fractions were present at higher concentrations and for longer periods of time in switchgrass and big bluestem than for smooth bromegrass. Although the three protein fractions were not identified, some mechanism apparently exists in warm-season species that allows certain protein fractions to remain undegraded for longer periods of time. Miller et al. (1996) offered three possibilities for the fate of warm-season grass forage protein (i) partial protection from digestion for up to 24 h, (ii) protein compartmentalized within the bundle sheath cell can exit the rumen and be hydrolyzed in the abomasum and small intestine, and (iii) complete protection of the protein and excretion from gastrointestinal tract. Thus, the ruminant animal's facilities for using the degraded protein may be increased due to fewer losses from ammonia and urinary excretions since the protein from warm-season grasses has a slower rate of degradation.

CONCLUSIONS

In ruminant production systems that use forages as the primary source of energy, purchasing protein supplements is the most costly input. Profitable animal performance can only be achieved when rate and extent of forage protein is balanced with rate and extent of organic matter digestibility. Greater use of available forage protein would enhance profitability and simplify development of protein supplementation schemes that may be system specific.

Recent advances in protein characterization will probably alter protein supplementation strategies and result in less reliance on CP estimates. The apparent differences in ruminal degradation of forage proteins in cool- and warm-season grasses suggest that current protein supplementation approaches should be altered. The direct causes for differences in response to supplemental protein for cattle grazing cool- and warm-season grasses, however, have not been determined. Since response to supplemental protein in grazing ruminants is highly variable and species specific, protein supplementation requirements on pasture will become more sophisticated as plant protein composition and utilization becomes defined by species.

REFERENCES

Aharoni, Y., and H. Tagari. 1991. Use of nitrogen-15 determinations of purine nitrogen fraction of digesta to define nitrogen metabolism traits in the rumen. J. Dairy Sci. 74:2540–2547.

Akin, D.E. 1980. Evaluation by electron microscopy and anaerobic culture of types of rumen bacteria associated with digestion of forage cell walls. Appl. Environ. Microbiol. 39:242–252.

Akin, D.E., and D. Burdick. 1975. Percentage of tissue types in tropical and temperate grass leaf blades and degradation of tissues by microorganisms. Crop Sci. 15:661–668.

Amos, H.E., D.R. Windham, and J.J. Evans. 1984. Nitrogen metabolism of steers fed suncured hay and drum dehydrated alfalfa and coastal bermudagrass. J. Anim. Sci. 58:987–995.

Anderson, S.J., T.J. Klopfenstein, and V.A. Wilkerson. 1988. Escape protein supplementation of yearling steers grazing smooth brome pastures. J. Anim. Sci. 66:237–242.

Annison, F.N., and D. Lewis. 1959. Metabolism in the rumen. Methuen & Co., London.

Barry, T.N., and T.R. Manley. 1984. The role of condensed tannins in the nutritional value of *Lotus pedunculatus* for sheep. Brit. J. Nutr. 51:493–504.

Beever, D.E., S.B. Cammell, and A.S. Wallace. 1974. The digestion of fresh, frozen and dried perennial ryegrass. Proc. Nutr. Soc. 33:73A–74A.

Beever, D.E., M.S. Dhanoa, H.R. Losada, R.T. Evans, S.B. Cammell, and J. France. 1986. The effect of forage species and stage of harvest on the process of digestion occurring in the rumen of cattle. Brit. J. Nutr. 56:439–454.

Beever, D.E., and R.C. Siddons. 1986. Digestion and metabolism in the grazing ruminant. p. 479–497. *In* L. P. Milligan et al. (ed.) Control of digestion and metabolism in ruminants. Prentice-Hall, Englewood Cliffs, NJ.

Beever, D.E., D.J. Thomson, E. Pheffer, and D.G. Armstrong. 1971. The effects of drying and ensiling grass on its digestion in sheep. Sites of energy and carbohydrate digestion. Brit. J. Nutr. 26:123–134.

Blasi, D.A., J.K. Ward, T.J. Klopfenstein, and R.A. Britton. 1991. Escape protein for beef cows: III. Performance of lactating beef cows grazing smooth brome or big bluestem. J. Anim. Sci. 69:2294–2302.

Brady, C.J. 1976. Plant proteins, their occurrence, quality, and distribution. p. 13–16. *In* T.H. Sutherland et al. (ed.) From plant to animal protein: Reviews in rural science. no. II. Univ. of New England Publ. Unit, Armidale.

Britton, R.A., D.P. Colling, and T.J. Klopfenstein. 1978. Effects of complexing sodium bentonite with soybean meal or urea in vitro ruminal ammonia release and nitrogen utilization in ruminants. J. Anim. Sci. 46:1738–1747.

Broderick, G.A. 1993. Quantifying forage protein quality. p. 200–228. *In* G.C. Fahey et al. (ed.) Forage quality, evaluation, and utilization. ASA, CSSA, and SSSA, Madison, WI.

Brown, R.H. 1978. A difference in N use efficiency in C_3 and C_4 plants and its implication in adaptive and evolution. Crop Sci. 18:93–98.

Bryant, M.P., and I.M. Robinson. 1963. Apparent incorporation of ammonia and amino acid carbon during growth of selected species of ruminal bacteria. J. Dairy Sci. 46:150–154.

Burroughs, W., D.K. Nelson, and D.R. Mertens. 1975. Protein physiology and its application in the lactating cow: The metabolizable protein feeding standard. J. Anim. Sci. 41:933–944.

Burroughs, W., A.H., Trenkle, and R.L. Vetter. 1974. Metabolizable protein (amino acid) feeding standard for cattle and sheep fed rations containing either alpha-amino or non-protein nitrogen, Iowa State Univ. Coop. Ext. Serv. A. S. Leaflet R190. Iowa State Univ. Press, Ames.

Buxton, D.R., D.R. Mertens, and D.S. Fisher. 1996. Forage quality and ruminant utilization. p. 229–266. *In* L.E. Moser et al. (ed.) Cool-season forage grasses. Agron. Monogr. 34. ASA, CSSA, and, SSSA, Madison, WI.

Chalupa, W. 1975. Rumen bypass and protection of proteins and amino acids. J. Dairy Sci. 58:1198–1218.

Cherney, D.J.R., J.J. Volenec, and J.H. Cherney. 1992. Protein solubility and degradation in vitro as influenced by buffer and maturity of alfalfa. Anim. Feed Sci. Technol. 37:9–20.

Craig, W.M., G.A. Broderick, and D.B. Ricker. 1987. Quantitation of microorganisms associated with the particulate phase of ruminal ingesta. J. Nutr. 117:56–62.

Cuomo, G.J., and B.E. Anderson. 1996. Nitrogen fertilization and burning effects on rumen protein degradation and nutritive value of native grasses. Agron. J. 88:439–442.

Dehority, B.A., and J.A. Grubb. 1980. Effects of short-term chilling of rumen contents on viable bacteria numbers. Appl. Environ. Microbiol. 39:376–381.

Duncan, C.W., I.D. Agrawala, C.F. Huffman, and R.W. Luecke. 1953. A quantitative study of rumen synthesis in the bovine on natural and purified rations. J. Nutr. 49:41–49.

Ferguson, W.S., and R.A. Terry. 1954. The fractionation of non-protein nitrogen of grassland herbage. J. Sci. Food. Agric. 5:515–524.

George, J.R., and D.E. Farnham. 1991. Drying method and temperature effects on ruminal escape protein. p. 188. *In* Agronomy abstracts. ASA, Madison, WI.

Hafley, J.L., B.E. Anderson, and T.J. Klopfenstein. 1993. Supplementation of growing cattle grazing warm-season grass with proteins of various ruminal degradabilities. J. Anim. Sci. 71:522–529.

Hogan, J.P., and J.R. Lindsay. 1979. The digestion of nitrogen associated with the plant cell wall in the stomach and small intestine of sheep. Aust. J. Agric. Res. 31:147–153.

Hollingsworth, K.J., T.J. Klopfenstein, and M.H. Sindt. 1993. Metabolizable protein of summer annuals for growing calves. J. Prod. Agric. 6:415–418.

Huffaker, R.C. 1982. Biochemistry and physiology of leaf proteins. p. 370–400. *In* D. Boulter and B. Parthier (ed.) Nucleic acids and proteins in plants. no. I. Springer-Verlag, Berlin.

Hunter, R.A., and B.D. Siebert. 1985. Utilization of low-quality roughage by *Bos taurus* and *Bos indicus* cattle: 1. Rumen digestion. Brit. J. Nutr. 53:637–648.

Jones, W.T., and J.L. Mangan. 1976. Large-scale isolation of fraction 1 leaf protein (18S) from lucerne (*Medicago sativa* L.). J. Agric. Sci. (Camb.) 86:495–501.

Krishnamoorthy, U., C.J. Sniffen, and P.J. Van Soest. 1982. Nitrogenous fractions in feedstuffs. J. Dairy Sci. 65:217–225.

Lamport, D.T.A., and D.H. Northcote. 1960. Hydroxyproline in primary cell walls of higher plants. Nature (London) 188:665–666.

Lexander, K., R. Carlsson, V. Schalen, A. Simonsson, and T. Lundborg. 1970. Quantities and quality of leaf protein concentrates from wild species and crop species grown under controlled conditions. Annu. Appl. Biol. 66:193–216.

Lyttleton, J.W. 1973. Proteins and nucleic acids. p. 63–103. *In* G.W. Butler and R.W. Bailey (ed.) Chemistry and biochemistry of herbage. Vol. 1. Academic Press, London.

Mackie, R.I., J.J. Therion, F.M.C. Gilchrist, and M. Ndhlovu. 1983. Processing ruminal ingesta to release bacteria attached to feed particles. S. Afr. J. Anim. Sci. 13:52–54.

Mangan, J.L. 1982. The nitrogenous constituents of fresh forage. p. 25–40. *In* D.J. Thomsom et al. (ed.) Forage protein in ruminant animal production. no. 6. Occ. Publ. BSAP, Edinburgh, England.

Mangan, J.L., R.L. Vetter, D.J. Jordan, and P.C. Wright. 1976. The effect of condensed tannins of sainfoin (*Onobrychis viciaefolia*) on the release of soluble leaf protein into the food bolus of cattle. Proc. Nutr. Soc. 35:95A–97A.

McMeniman, N.P., I.F. Beale, and K.M. Murphy. 1986. Nutritional evaluation of southwest Queensland pastures: II. The intake and digestion of organic matter and nitrogen by sheep grazing on Mitchell grass and mulga/grassland associations. Aust. J. Agric. Res. 37:303–314.

Mehrez, A.Z., and E.R. Orskov. 1977. A study of the artificial fibre bag technique for determining the degradability of feeds in the rumen. J. Agric. Sci. (Camb.) 88:645–650.

Merry, R.J., and A.B. McAllen. 1983. A comparison of the chemical composition of mixed bacteria harvested from the liquid and solid fractions of rumen digesta. Brit. J. Nutr. 50:701–709.

Miller, M.S., L.E. Moser, S.S. Waller, T.J. Klopfenstein, and B.H. Kirch. 1996. Immunofluorescent localization of RuBPCase in degraded C_4 grass tissue. Crop Sci. 36:169–175.

Minato, H., A. Endo, M. Hiuchi, Y. Ootomo, and T. Uemura. 1966. Ecological treatise on the rumen fermentation: I. The fractionation of bacteria attached to the rumen digesta solids. J. Gen. Appl. Microbiol. 12:39–52.

Minson, D.J. 1990. Forage in ruminant nutrition. Academic Press, New York.

Mitchell, R.B., R.A. Masters, S.S. Waller, K.J. Moore, and L.E. Moser. 1994. Big bluestem production and forage quality responses to burning and fertilizer in tallgrass prairies. J. Prod. Agric. 7:355–359.

Mitchell, R.B., D.D. Redfearn, L.E. Moser, R.J. Grant, K.J. Moore, and B.H. Kirch. 1997. Relationships between in situ protein degradability and grass developmental morphology. J. Dairy Sci. 80:1143–1149.

Mullahey, J.J., S.S. Waller, K.J. Moore, L.E. Moser, and T.J. Klopfenstein. 1992. In situ ruminal protein degradation of switchgrass and smooth bromegrass. Agron. J. 84:183–188.

Muscato, T.V., C.J. Sniffen, U. Krishnamoorthy, and P.J. Van Soest. 1983. Amino acid content of non-cell and cell wall fractions in feedstuffs. J. Dairy Sci. 66:2198–2207.

Nocek, J., and J.B. Russell. 1988. Protein and carbohydrate as an integrated system. Relationship of ruminal availability to microbial contribution and milk production. J. Dairy Sci. 71:2070–2107.

National Research Council. 1985. Ruminant nitrogen usage. Natl. Academy Press, Washington, DC.

Owens, F.N., and R.A. Zinn. 1988. Protein metabolism of ruminant animals. p. 227–249. *In* D.C. Church (ed.) The ruminant animal. Prentice-Hall, Englewood Cliffs, NJ.

Poos-Floyd, M., T. Klopfenstein, and R.A. Britton. 1985. Evaluation of laboratory techniques for predicting ruminal protein degradation. J. Dairy Sci. 68:829–839.

Redfearn, D.D., L.E. Moser, S.S. Waller, and T.J. Klopfenstein. 1995. Ruminal degradation of switchgrass, big bluestem, and smooth bromegrass leaf proteins. J. Anim. Sci. 73:598–605.

Russell, J.B., J.D. O'Connor, D.G. Fox, P.J. Van Soest, and C.J. Sniffen. 1992. A net carbohydrate and protein system for evaluating cattle diets: I. Ruminal fermentation. J. Anim. Sci. 70:3551–3561.

Salisbury, F.B., and C. W. Ross. 1985. Plant physiology. 3rd ed. Wadsworth Publ. Co., Belmont, CA.

Sanderson, M.A., and W.F. Wedin. 1989a. Nitrogen concentrations in the cell wall and lignocellulose of smooth bromegrass herbage. Grass For. Sci. 44:151–158.

Sanderson, M.A., and W.F. Wedin. 1989b. Nitrogen in the detergent fibre fractions of temperate legumes and grasses. Grass For. Sci. 44:159–168.

Sniffen, C.J., J.D. O'Connor, P.J. Van Soest, D.G. Fox, and J.B. Russell. 1992. A net carbohydrate and protein system for evaluating cattle diets: II. Carbohydrate and protein availability. J. Anim. Sci. 70:3562–3577.

Steinhour, W.D., and J.H. Clark. 1980. Microbial nitrogen flow to the small intestine of ruminants. p. 166–182. *In* F.N. Owens (ed.) Proc. Int Symp., Stillwater, OK. 19–21 Nov. 1980. Oklahoma State Univ., Stillwater.

Stern, M.D., and L.D. Satter. 1980. In vivo estimation of protein degradability in the rumen. p. 57–71. *In* F.N. Owens (ed.) Proc. of Int. Symp., Stillwater, OK. 19–21 Nov. 1980. Oklahoma State Univ., Stillwater.

Stevenson, I.L. 1978. The production of extracellular amino acids by rumen bacteria. Can. J. Microbiol. 24:1236–1241.

Tamminga, S. 1979. Protein degradation in the forestomachs of ruminants. J. Anim. Sci. 49:1615–1630.

Tilley, J.M.A., and R.A. Terry. 1963. A two-stage technique for the in vitro digestion of forage crops. J. Br. Grassl. Soc. 18:104–111.

Uedan, K., and T. Sugiyama. 1976. Purification and characterization of phosphoenolpyruvate carboxylase from maize leaves. Plant Physiol. 57:906–910.

Van Soest, P.J. 1982. Nutritional Ecology of the Ruminant. O & B Books, Corvallis, OR.

Walker, D.J., A.R. Egan, C.J. Nader, M.J. Ulyatt, and G.B. Storer. 1975. Rumen microbial protein synthesis and proportions of microbial and non-microbial nitrogen flowing to the intestines on sheep. Aust. J. Agric. Res. 26:699–708.

Wanderley, R.C., J.T. Huber, Z. Wu, M. Pessarakli, and C. Fontes, Jr. 1993. Influence of microbial colonization of feed particles on determination of nitrogen degradability by in situ incubation. J. Anim. Sci. 71:3073–3077.

Weiss, W.P., H.R. Conrad, and W.L. Shockey. 1986. Digestibility of nitrogen in heat-damaged alfalfa. J. Dairy Sci. 69:2658–2670.

Wilkerson, V.A., T.J. Klopfenstein, and W.W. Stroup. 1995. A collaborative study of in situ forage protein degradation. J. Anim. Sci. 73:583–588.

Zinn, R.A., and F.N. Owens. 1986. A rapid procedure for purine measurement and its use for estimating net ruminal protein synthesis. J. Anim. Sci. 66:157–166.

2 Fiber Composition and Digestion of Warm-Season Grasses

Kenneth J. Moore and Dwayne R. Buxton
Iowa State University
Ames, Iowa

The nutritive value of warm-season grasses is determined largely by the amount of digestible energy (DE) that the animal can consume from them (Moore, 1980; Gillet et al., 1985). When livestock consume diets with a high proportion of warm-season grass, DE intake is limited primarily by the high concentration and low digestibility of fiber in the grass. This relatively high concentration of fiber in warm-season grasses often has an overriding influence on both dry matter digestibility and intake (Abrams et al., 1983; Mertens, 1987, 1993).

Plant cell walls are the major source of dietary fiber for animals. Polysaccharides in cell walls cannot be degraded by mammalian enzymes so animals depend on microbial fermentation to degrade them. Ruminants are especially well-adapted for using plant fiber for energy (Chesson & Forsberg, 1988; Van Soest et al., 1991). Fiber, measured as neutral detergent fiber (NDF), usually accounts for 30 to 80% of the organic matter in forage crops. The remaining organic matter is collectively referred to as cell solubles and is almost completely digestible. The nutritional availability of fiber to livestock, however, varies greatly depending on its composition and structure. Lignin has been identified as primarily responsible for limiting digestibility of fiber, but fiber use also is limited by physical constraints at the cellular organization level (Wilson & Mertens, 1995).

FIBER COMPOSITION

Cellulose, hemicelluloses, and lignin are the primary chemical constituents that comprise plant fiber (Moore & Hatfield, 1994). Cellulose is a linear polymer of β-D-glucose units linked by (1→4) glycosidic bonds. Many of the unique properties of cellulose arise from its secondary structure. Linear chains of glucose units combine to form microfibrils, which are extensively cross-linked by hydrogen bonding. Microfibrils are insoluble, hydrophobic, and more resistant to digestion than the glucan chains from which they are formed (Cowling, 1975; Kerley et al., 1988).

Hemicelluoses consist of complex heteropolymers that vary considerably in primary composition, substitutions, and degree of branching. Arabinoxylans and

---

Copyright © 2000. Crop Science Society of America and American Society of Agronomy, 677 S. Segoe Rd., Madison, WI 53711, USA. *Native Warm-Season Grasses: Research Trends and Issues.* CSSA Special Publication no. 30.

Table 2–1. Ranges in composition and digestibility of some common forage grasses. Data adapted from an experiment by Jung and Vogel (1986).

Species	Concentration				Digestibility		
	Fiber	Cellulose	Hemi-cellulose	Lignin	Fiber	Cellulose	Hemi-cellulose
	———— g kg^{-1} DM ————				——— g kg^{-1} ———		
Cool-Season Grasses							
Orchardgrass	431–608	174–318	240–261	35–40	549–859	572–876	618–909
Bromegrass	382–645	187–321	178–270	15–49	381–789	375–786	468–886
Warm-Season Grasses							
Switchgrass	577–743	242–376	308–316	20–60	238–793	242–799	307–828
Indiangrass	697–759	314–378	327–351	25–51	373–681	366–676	412–726
Big Bluestem	674–729	321–356	319–326	30–46	300–556	300–563	341–601

xyloglucans are the predominant polysaccharides comprising hemicelluloses of grasses (Wilkie, 1979; Chesson & Forsberg, 1988). Arabinoxylans are substituted with glucuronic acid and 4-O-methylgucuronic acid to varying degrees forming glucuronoarabinoxylans. Arabinoxylans also may be substituted with galactose and acetyl groups (Hatfield, 1989).

Lignin is a highly-condensed polymer formed by the dehydrogenative polymerization of the hydroxycinnamyl alcohols, p-coumaryl, coniferyl, and sinapyl alcohols (Sarkanen & Ludwig, 1971; Dean & Eriksson, 1992). It is considered virtually indigestible (Van Soest, 1994) by ruminants and is implicated in limiting the digestion of cell-wall polysaccharides (Jung, 1989). The close physical association of lignin with the cell-wall polysaccharides and the existence of covalent bonds between lignin and cell-wall polysaccharides are believed to be the major factors limiting the accessibility of the polysaccharides as substrate for the hydrolases secreted by rumen microbes (Chesson, 1982, 1993; Jung & Deetz, 1993; Jung & Allen, 1995).

Warm-season perennial grasses generally have higher structural carbohydrate concentrations than cool-season grasses (Reid et al., 1988; Akin & Chesson 1989; Buxton & Casler, 1993; Kephart & Buxton, 1993; Buxton & Fales, 1994; Buxton et al., 1995). Jung and Vogel (1986) conducted an experiment in which they examined the relationship between lignin concentration and digestibility in several grass species common to the Central Great Plains. Grasses were sampled at several stages of maturity and concentrations and digestibility of fiber constituents were determined (Table 2–1). Warm-season species had higher concentrations of fiber and fiber constituents than cool-season grasses. When expressed as a proportion of total fiber, cellulose and hemicellulose concentrations, however, were similar between cool- and warm-season grasses. Therefore, even though warm-season grasses had higher concentrations of fiber than cool-season species, the composition of their fiber composition was similar.

Hydrolysis of grass hemicelluloses yields the neutral monosaccharides glucose, xylose, arabinose, mannose and galactose, and the uronic acids, galaturonic, glucuronic acid, and 4-O-methylgucuronic acid (Wilkie, 1979; Åman & Graham, 1990). The relative proportions of each monosaccharide varies among species, re-

Table 2–2. Proportion of various tissues in stem, leaf blade, and leaf sheath cross sections of switchgrass. Data from Twidwell et al. (1991).

Tissue	Stem	Leaf blade	Leaf sheath
		%	
Epidermis	3.3	23.8	9.3
Sclerenchyma	--	2.9	9.3
Mesophyll	--	35.7	--
Parenchyma	65.2	--	44.9
Bundle sheath	6.8	24.6	7.4
Cortex	10.9	--	--
Phloem	1.5	1.4	1.8
Xylem	8.1	2.9	4.7

flecting differences in polysaccharide structures. Xylose and arabinose account for most of the neutral sugars in hemicelluloses isolated from grasses (Collings & Yokoyama, 1979; Wedig et al., 1987). Buxton et al. (1987) conducted comparative studies on structural neutral sugars isolated from stem bases of several grasses. Glucose (predominantly of cellulose origin) and xylose accounted for 62 and 30% of the neutral sugars hydrolyzed, respectively. Arabinose concentrations averaged 4.4 %. There seems to be no clear distinction between the neutral sugar composition of structural carbohydrates of warm and cool-season grasses. Windham et al. (1983) found similar concentrations of hemicellulosic neutral sugars among three warm and three cool-season grasses when adjusted for recoveries.

Stems of most forage grasses have a greater fiber concentration than do leaf blades (Griffin & Jung, 1983; Jung & Vogel, 1992). Leaf sheaths, the portion of the leaf enclosing the stem of grasses, are usually intermediate to leaf blades and stems in fiber concentration (Twidwell et al., 1988). Higher fiber concentrations in stems occur in part because stems contain more structural and conducting tissues than leaves, whereas a larger proportion of leaves is occupied by thin-walled mesophyll cells (Akin & Chesson, 1989; Wilson, 1993). An analysis of anatomical tissue in switchgrass (*Panicum virgatum* L.) (Twidwell et al., 1991) indicated that stems contained a higher proportion of less degradable tissues than did either leaf blades or sheaths (Table 2–2).

Jung and Vogel (1992) determined concentrations of fiber constituents in leaves and stems of switchgrass and big bluestem [*Andropogon gerardii* (L.) Vitman] at three stages of maturity. Across maturities, total fiber concentrations averaged 61 and 42 g kg^{-1} DM higher in stems than leaves of switchgrass and big bluestem, respectively (Fig. 2–1). Most of the difference in total fiber concentration between leaves and stems was accounted for by higher cellulose and lignin concentrations in stems. No difference in fiber concentration of stems was observed between species, but fiber concentration was higher in leaves of big bluestem than switchgrass at all stages of maturity. Lignin concentration increased with maturity in both leaves and stems.

In another study, Twidwell et al. (1988) evaluated the influence of plant morphology on forage quality of three switchgrass cultivars. 'Cave-in-Rock' had a lower fiber concentration than 'Pathfinder' and 'Trailblazer'. Fiber concentration was highest for stems (802 g kg^{-1} DM), lowest for leaf blades (621 g kg^{-1} DM), and intermediate for leaf sheaths (737 g kg^{-1} DM). Hemicellulose concentrations were

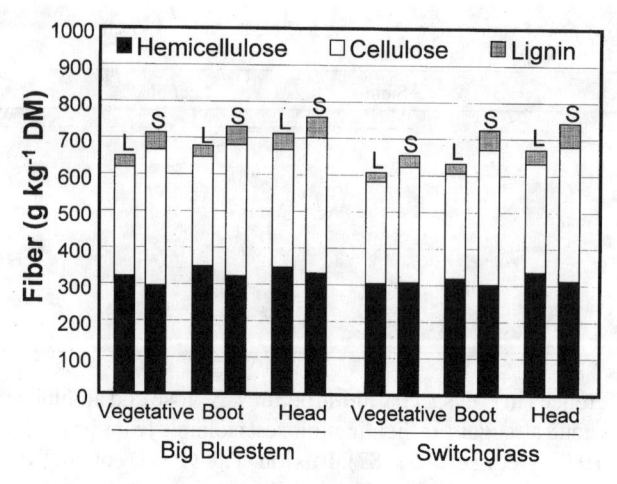

Fig. 2–1. Concentrations of fiber constituents in leaves (L) and stems (S) of big bluestem and switchgrass at three stages of maturity (data from Jung & Vogel, 1992).

similar among plant parts; however, concentrations of lignin and cellulose were higher in stems than in either leaf blades or sheaths.

Fiber concentration increases as grasses mature mostly due to changes in their morphology (leaf/stem ratio) and to a lesser degree, aging of plant organs (Buxton & Fales, 1994; Buxton et al., 1995). Griffin and Jung (1983) observed that the proportion of leaf tissue in dry matter of switchgrass and big bluestem on average declined from 68 to 30% with maturation. The percentage of leaf tissue, however, was generally higher in switchgrass than big bluestem at all sampling dates. They noted that concentrations of fiber in leaves changed little with respect to maturity, but increased rapidly in stems. Similar effects of plant maturity on fiber concentration of switchgrass were observed by Twidwell et al. (1988).

Aging of individual plant organs has a relatively smaller effect on fiber concentration as compared to morphological development. Anderson (1985) studied the aging of switchgrass leaves as they developed over time. He reported that neutral detergent fiber concentration of leaves increased an average of 1.3 g kg^{-1} DM d^{-1} from the second-leaf stage into midsummer. He also noted that leaves that developed early in the season had lower fiber concentrations than those that developed later. This difference was attributed to warmer growing temperatures later in the season.

Mitchell et al. (1999) studied the relationship between morphological development and nutritive value of switchgrass and big bluestem. Fiber concentration of both species increased nonlinearly with respect to mean stage count (MSC) and was lower for switchgrass at any given MSC (Fig. 2–2). The MSC represents the mean developmental stage of a population of grass tillers (Moore et al., 1991; Mitchell et al., 1997). In both species, the rate of change in fiber concentration began to decrease during the elongative stages (MSC 2.0–2.9) and was nearly asymptotic when the mean stage of the tiller population approached the reproductive stages (MSC 3.0–3.9).

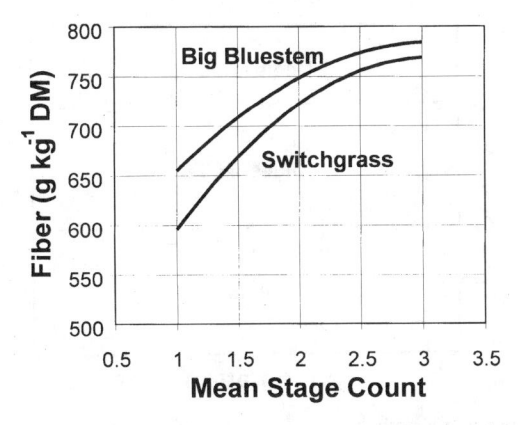

Fig. 2–2. Fiber concentrations of big bluestem and switchgrass as a function of mean stage count. Calculated from equations published by Mitchell et al., 2000.

FIBER DIGESTION

True digestibility (TD) of a forage is the actual proportion of dry matter that is digested exclusive of any endogenous contributions of the animal or rumen microbes to the undigested residue. It represents the availability of nutrients present in the forage to the animal consuming them and is closely related to DE. True digestibility can be determined by subtracting undigested fiber from total dry matter intake (DMI) and expressing this quantity as a proportion of DMI. This calculation is based on the assumption that all constituents of the diet other than fiber are completely digested (Van Soest, 1994). Therefore, TD can be partitioned into two components: cell solubles (C_S) and digestible fiber (Moore et al., 1993). Cell solubles includes all dry matter constituents other than fiber and is considered completely digestible.

Digestion of fiber is a time-dependent process. Degradation of plant fiber in situ, as well as in in vitro batch culture, can be described by a first-order kinetic model (Moore & Cherney, 1986). The model, however, must be modified to account for a discrete lag time and to exclude the indigestible portion of the fiber (Mertens, 1993). According to the model, the extent of fiber digestion in the rumen is dependent upon the rate of fiber digestion (k), the concentration of potentially digestible fiber (C_D), digestion lag time (L), and the residence time in the rumen (Mertens, 1993). If these parameters are known, TD can be calculated as: TD = C_S + $C_D(1 - e^{-k(t-L)})$ where $(1 - e^{-k(t-L)})$ represents the digestion coefficient for fiber (Fig. 2–3) (Moore et al., 1993).

Residence time in the rumen is a reciprocal function of the rate of passage from the rumen and determines the length of time fiber will be exposed to rumen fermentation. Long retention of slowly digesting particles in the rumen results in rumen fill and reduced intake of additional forage. Simulations with mathematical models suggest fiber concentration of plants, extent of fiber digestion, and rate of passage, including rate of particle size reduction, are the most important factors affecting particle retention in the rumen and intake (Mertens, 1993). Rate of fiber digestion and length of lag time before digestion begins have less effect on intake.

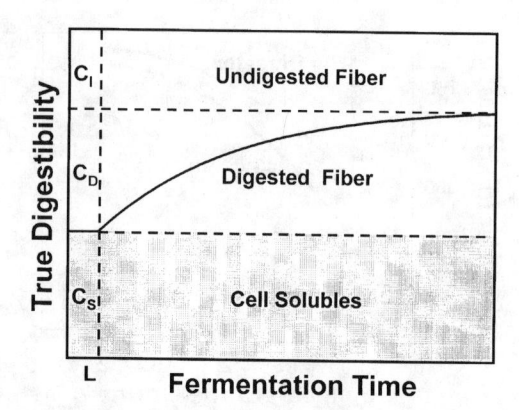

Fig. 2–3. Components of true digestibility (TD) according to the model TD = $C_S + C_D(1 - e^{-k(t-L)})$ where: C_S is the concentration of cell solubles, C_D is the concentration of digestible fiber, k is the rate of fiber digestion, L is the digestion lag time, and t is the residence time in the rumen (from Moore et al., 1993).

Fiber concentration often is related negatively to the fractional rate constant of digestion (Smith et al., 1972; Fisher et al., 1989). Forages with high fiber concentrations may have walls that are thickened, and therefore, more resistant to mechanical disruption and microbial penetration and digestion. Available surface area for microbial attack may limit rate of digestion in these situations. Parenchyma cells usually are degraded rapidly and extensively in situ, which may occur, in part, because of their small size, thin cell walls, and large surface area exposed to rumen microorganisms after mastication. On the other hand, sclerenchyma cells are slowly and incompletely degraded in situ, which may occur in part because of their large size, thick cell walls, and low surface area exposed to digestion after mastication (Grabber et al., 1992). Longer lag times also have been associated with the lower accessibility of forage cell walls to microorganisms. Attachment of bacteria to cells is necessary for digestion to occur and may be a partial cause of lag time (Allen & Mertens, 1988).

Digesta particles must be reduced to a size that will pass through a 1-mm screen before they can easily flow from the rumen (Minson, 1990). Leaves usually are retained in the rumen for a shorter time than stems because both passage rate and fiber digestion is faster for leaves. Plant factors that affect rate and extent of cell-wall digestion may also influence fragility (brittleness) and cell-wall disintegration in the rumen.

In addition to particle size, particle density plays a role in passage of particles from the rumen. Specific gravity of ruminal contents is lowest after eating and approaches unity before the next meal. When particles begin to ferment rapidly, gasses produced by fermentation can buoy them keeping their specific gravity to <1.2. Once digestion of particles slows as it nears completion, and particles are completely hydrated, they achieve a specific gravity of between 1.2 and 1.4, which is optimal for passage from the rumen (Wattiaux et al., 1992).

Digestibility of warm-season grasses generally is lower than that of cool-season grasses at similar stages of maturity. Minson (1990) reported that the average digestibility of cool-season grasses was 13 g kg^{-1} greater than that of warm-sea-

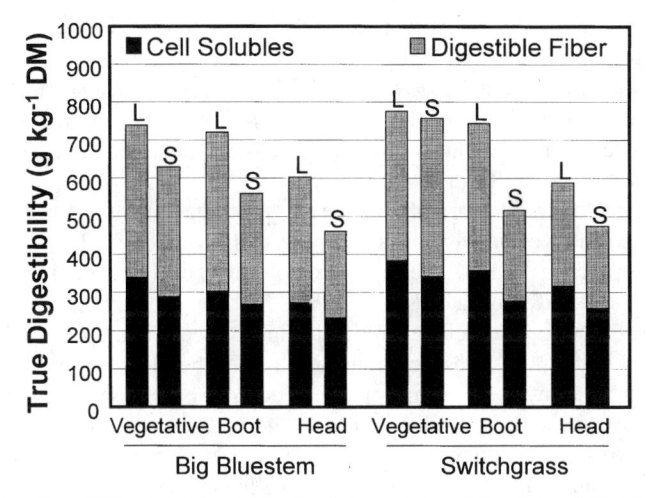

Fig. 2–4. True digestibility of big bluestem and switchgrass leaves (L) and stems (S) at three stages of maturity (data from Jung & Vogel, 1992).

son grasses. Reid et al. (1988) found that the average digestibility of warm-season grasses was lower than that of cool-season grasses. Kephart and Buxton (1993), however, found no consistent differences between the warm- and cool-season species for DM digestibility. Cool-season grasses often have a lower proportion of stems in herbage than do warm-season grasses, which may account for much of the lower forage quality of warm-season grasses.

When plants advance in maturity, the leaf/stem ratio usually decreases. Additionally, fiber concentration within stems and usually within leaves increases, and the proportion of cell solubles decreases. As maturity advances, grasses develop more lignified cells, which contribute to the high cell-wall concentration and low digestibility of grass leaves (Buxton & Casler, 1993).

Digestibility of grass leaves generally is higher than that of stems. Jung and Vogel (1992) determined the digestibility of fiber constituents in leaves and stems of switchgrass and big bluestem at three stages of maturity (Fig. 2–4). Across maturities they observed no differences in fiber digestibility of leaves between species. At the vegetative stage, however, switchgrass stems had a higher fiber digestibility than did big bluestem. True digestibility declined with maturity in both species and was generally higher for leaves than stems with the exception of the vegetative stage (Fig. 4).

Twidwell et al. (1988) compared the digestibility of three switchgrass cultivars at three stages of maturity. Averaged across plant parts, extent of fiber digestion decreased from 660 g kg^{-1} at a late vegetative stage to 540 g kg^{-1} 28 d later. Extent of fiber digestion for leaf blades was 740 g kg^{-1} averaged across maturities compared with 430 and 590 g kg^{-1} for stems and leaf sheaths, respectively. Rate of digestion (k) of leaf blades and sheaths were not affected by maturity, but k of stems decreased with maturity from an average of 0.065 to 0.041 h^{-1}.

Leaf blades of cool-season grasses usually are more digestible than those of warm-season grasses because they have more mesophyll cells. In cool-season species, mesophyll, phloem, epidermal and parenchyma bundle sheath cells are all totally degraded. In contrast, the epidermal and parenchyma bundle sheath cells in warm-season grasses may be slowly or only partially degraded (Akin, 1989). The greater fiber concentration in warm-season compared with cool-season grasses is primarily due to larger amounts of vascular tissue and parenchyma bundle sheath cells (Akin & Burdick, 1975).

Ruminal microbes normally gain access for attachment to fiber through cut or sheared surfaces of herbage because epidermal layers of plants restrict microbe and enzyme access. Plant tissue types are colonized and degraded by rumen microorganisms at different rates (Akin, 1989). Mesophyll and phloem cells are degraded more rapidly and before the outer bundle sheath and epidermal cells. Attachment of bacteria to sclerenchyma and vascular bundle cells is low and these tissues degrade slowly and incompletely. Cool-season grasses have leaves with widely spaced vascular bundles and less distinct parenchyma bundle sheaths, whereas warm-season grasses have leaves with Kranz anatomy resulting in closely spaced vascular bundles and a distinct thick-walled parenchyma bundle sheath surrounding each bundle. Cool-season grass leaves have a higher proportion of loosely arranged mesophyll cells than warm-season grasses, which are usually the first to be digested. The proportions of epidermis and parenchyma bundle sheath are substantially higher in warm-season grass leaf blades than those of cool-season grasses. These tissues usually are slowly or partially degraded in warm-season grasses, but extensively degraded in cool-season grasses. In contrast to leaf anatomy, stem anatomy of warm- and cool-season grasses is similar as is the extent of digestibility (Akin & Chesson, 1989; Wilson, 1993).

Anatomical structure of grass stems also may influence their rate of passage. Lignification of parenchyma in grass stems nearing maturity results in a compact, interlocked residue after digestion, comprised of epidermis, sclerenchyma ring, vascular tissues, and parenchyma. The relatively long, filamentous shape of grass particles may result in a greater resistance to escape from the rumen than the cuboidal fragments of legumes (Mertens et al., 1984; Wilson, 1993).

Physical and structural barriers may limit fiber digestibility beyond the effect of lignin (Brazle et al., 1979). Waxes and the cuticle of the epidermis covering plants restrict microbe and enzyme access to forage tissues (Wilson & Kennedy, 1996). Ruminal microbes normally enter interior cells in leaf blades and stems through stomata, fractures in the cuticle, or through cut or sheared surfaces. The epidermis of cool-season grasses is linked to thin-walled parenchyma cells, whereas in warm-season grasses the linkage is to thick-walled bundle sheath cells. The epidermis is usually shed within 24 h during digestion of cool-season grasses, but may remain attached for longer than 48 h in warm-season grasses (Akin, 1989; Wilson, 1993). Grass stems have a ring of thick-walled, lignified cells resistant to digestion (Akin, 1989).

SUMMARY AND CONCLUSIONS

Fiber composition and digestibility are important factors limiting the nutritive value of warm-season grasses. Compared with other grasses at similar maturity, warm-season grasses generally have higher concentrations of fiber that is often lower in digestibility. Therefore, warm-season grasses have lower concentrations cell solubles and digestible fiber resulting in a lower digestible energy value.

As warm-season grasses mature, fiber concentration increases and fiber digestibility decreases. Changes in fiber concentration and digestibility with maturity mostly are related to changes in plant morphology. As grasses mature, the ratio of leaves to stems decreases. Stems have higher proportions of structural tissues that are high in fiber and often lignified. In addition to the negative effect of lignin per se on digestibility, the ultrastructure of these tissues can limit access by rumen microbes and therefore reduce digestibility.

Strategies for improving the digestible energy value of warm-season grasses should minimize the accumulation of indigestible fiber in the plant (Mertens & Ely, 1983; Moore et al., 1993). This may be achieved by decreasing fiber concentration thereby increasing cell solubles, increasing fiber digestibility, or both. From a practical perspective this can be accomplished best by using warm-season grasses when they are in a vegetative stage of maturity. Good management of warm-season grass swards can greatly reduce the negative impacts of fiber on animal performance.

There are differences among warm-season grass species in fiber concentration and digestibility. For example, switchgrass typically has a lower fiber concentration than big bluestem at similar stages of maturity (Mitchell et al., 2000). Because switchgrass matures earlier than big bluestem, however, the difference is not generally recognized. By understanding differences in quality and maturity among warm-season grasses it should be possible to develop systems for improved utilization that exploit these differences.

Even within warm-season grass species there are inherent differences in fiber concentration and digestibility. Cultivars of switchgrass have been developed for improved DM digestibility (Vogel et al., 1991, 1996). In general, improvements in switchgrass digestibility have been accomplished by increasing fiber digestibility with little or no change in fiber concentration (Moore et al., 1993). There appears to be sufficient genetic variation in fiber concentration of some switchgrass populations to make progress in increasing true digestibility by breeding for lower fiber concentration (Godshalk et al., 1988a,b).

REFERENCES

Abrams, S.M., H. Hartadi, C.M. Chaves, J.E. Moore, and W.R. Ocumpaugh. 1983. Relationship of forage evaluation techniques to the intake and digestibility of tropical grasses. p. 508–511. In J.A. Smith and V.W. Hays (ed.) Proc. XIV Int. Grassl. Cong., Lexington, KY. 15–24 June 1981. Westview Press, Boulder, CO.

Akin, D.E. 1989. Histological and physical factors affecting digestibility of forages. Agron. J. 81:17–25.

Akin, D.E., and D. Burdick. 1975. Percentage of tissue types in tropical and temperate grass leaf blades and degradation of tissues by rumen microorganisms. Crop Sci. 15:661–668.

Akin, D.E., and A. Chesson. 1989. Lignification as the major factor limiting forage feeding value especially in warm conditions. p. 1753–1760. In R. Desroches (ed.) Proc. XVI Int. Grassl. Congr., Nice, France. 4–11 Oct. 1989. Assoc. Francaise pour la Production Fourragere, Nice.

Allen, M.S., and D.R. Mertens. 1988. Evaluating constraints on fiber digestion by rumen microbes. J. Nutr. 118:261–27.

Åman, P., and H. Graham. 1990. Chemical evaluation of polysaccharides in animal feeds. p. 161–177. In J. Wiseman and D.J.A. Cole (ed.) Feedstuff evaluation. Butterworths, London.

Anderson, B. The influence of aging on forage quality of individual switchgrass leaves and stems. p. 947–949. *In* T. Yoshiyama (ed.) Proc. XV Int. Grassl. Cong., Kyoto, Japan. 14–31 Aug. 1985. Science Council of Japan, Nishi-nasumo, Tochigi-ken.

Brazle, F.K., L.H. Harbers, and C.E. Owensby. 1979. Structural inhibitors of big and little bluestem digestion observed by scanning electron microscopy. J. Anim. Sci. 48:1457–1463.

Buxton, D.R., and M.D. Casler. 1993. Environmental and genetic effects on cell wall composition and digestibility. p. 685–714. *In* H.G. Jung et al. (ed.) Forage cell wall structure and digestibility. ASA, CSSA, and SSSA, Madison, WI.

Buxton, D.R., and S.L. Fales. 1994. Plant environment and quality. p. 155–199. *In* G.C. Fahey (ed.) Forage quality, evaluation, and utilization. ASA, CSSA, and SSSA, Madison, WI.

Buxton, D.R., D.R. Mertens, and K.J. Moore. 1995. Forage quality for ruminants: Plant and animal considerations. Prof. Anim. Sci. 11:121–131.

Buxton, D.R., J.R. Russell, and W.F. Wedin. 1987. Structural neutral sugars in legume and grass stems in relation to digestibility. Crop Sci. 27:1279–1285.

Chesson, A. 1982. A holistic approach to plant cell wall structure and degradation. p. 85–90. *In* G. Wallace and L. Bell (ed.) Fibre in human and animal nutrition. The Royal Soc. of New Zealand, Wellington, NZ.

Chesson, A. 1993. Mechanistic models of forage cell wall degradation. p. 347–376. *In* H.G. Jung, D.R. Buxton, R.D. Hatfield, and J. Ralph (ed.) Forage cell wall structure and digestibility. ASA, CSSA, and SSSA, Madison, WI.

Chesson, A., and C.W. Forsberg. 1988. Polysaccharide degradation by rumen microorganisms. p. 251–284. *In* P. N. Hobson (ed.) The rumen microbial system. Elsevier Applied Sci., New York.

Collings, G.F., and M.T. Yokoyama. 1979. Analysis of fiber components in feeds and forages using gas-liquid chromatography. J. Agric. Food Chem. 27:373–377.

Cowling, E.B. 1975. Physical and chemical constraints in the hydrolysis of cellulose and lignocellulosic materials. Biotechnol. Bioeng. Symp. 5:163–181.

Dean, J.F.D., and K.E. Eriksson. 1992. Biotechnological modification of lignin structure and composition in forest trees. Holzforschung 46:135–147.

Fisher, D.S., J.C. Burns, and K.R. Pond. 1989. Kinetics of in vitro cell-wall disappearance and in vivo digestion. Agron. J. 81:25–33.

Gillet, M., J. Schehovic, and M. Lila. 1985. Research for selection criteria for grass quality, based upon animal references. p. 957–958. *In* T. Yoshiyama (ed.) Proc. XV Int. Grassl. Cong., Kyoto, Japan. 14–31 Aug. 1985. Science Council of Japan, Nishi-nasumo, Tochigi-ken.

Godshalk, E.B., W.F. McClure, J.C. Burns, D.H. Timothy, and D.S. Fisher. 1988a. Heritabilty of cell wall carbohydrates in switchgrass. Crop Sci. 28:736–742.

Godshalk, E.B., D.H. Timothy, and J.C. Burns. 1988b. Effectiveness of index selection for switchgrass forage yield and quality. Crop Sci. 28:825–830.

Grabber, J.H., G.A. Jung, S.M. Abrams, and D.B. Howard. 1992. Digestion kinetics of parenchyma and sclerenchyma cell walls isolated from orchardgrass and switchgrass. Crop Sci. 32:806–810.

Griffin, J.L., and G.A. Jung. 1983. Leaf and stem forage quality of big bluestem and switchgrass. Agron. J. 75:723–726.

Hatfield, R.D. 1989. Structural polysaccharides in forages and their degradability. Agron. J. 81:39–46.

Jung, H.G. 1989. Forage lignins and their effects on fiber digestibility. Agron. J. 81:33–38.

Jung, H.G., and M.S. Allen. 1995. Characteristics of plant cell walls affecting intake and digestibility of forages by ruminants. J. Anim. Sci. 73:2774–2790.

Jung, H.G., and D.A. Deetz. 1993. Cell wall lignification and degradability. p. 315–346. *In* H.G. Jung et al. (ed.) Forage cell wall structure and digestibility. ASA, CSSA, and SSSA, Madison, WI.

Jung, H.G., and K.P. Vogel. 1986. Influence of lignin on digestibility of forage cell wall material. J. Anim. Sci. 62:1703–1712.

Jung, H.G., and K.P. Vogel. 1992. Lignification of switchgrass (*Panicum virgatum*) and big bluestem (*Andropogon gerardii*) plant parts during maturation and its effect on fiber degradability. J. Sci. Food Agric. 59:169–176.

Kephart, K.D., and D.R. Buxton. 1993. Forage quality responses of C_3 and C_4 perennial grasses to shade. Crop Sci. 33:831–837.

Kerley, M.S., G.C. Fahey, Jr., J.M. Gould, and E.L. Lannotti. 1988. Effects of lignification, cellulose crystallinity and enzyme accessible space on the digestibility of plant cell wall carbohydrates by the ruminant. Food Microstructure 7:59–65.

Mertens, D.R. 1987. Predicting intake and digestibility using mathematical models of ruminal function. J. Anim. Sci. 64:1548–1558.

Mertens, D.R. 1993. Kinetics of cell wall digestion and passage in ruminants. p. 535–570. *In* H. G. Jung et al. (ed.) Forage cell wall structure and digestibility. ASA, CSSA, and SSSA, Madison, WI.

Mertens, D.R., and L.O. Ely. 1983. Using a dynamic model of fiber digestion and passage to evaluate forage quality. p. 505–508. *In* J. Allan Smith and V.W. Hays. (ed.) Proc. XIV Int. Grassland Congr., Lexington, KY. 15–24 June 1981. Westview Press, Boulder, CO.

Mertens, D.R., T.L. Strawn, and R.S. Cardoza. 1984. Modelling ruminal particle size reduction: Its relationship to particle size description. p. 134–141. *In* P.M. Kennedy (ed.) Techniques in particle size analysis of feed and digesta in ruminants. Can. Soc. Anim. Sci. Occ. Publ. no. 1. Can. Soc. Anim. Sci., Edmonton, Alberta, Canada.

Minson, D.J. 1990. Forage in ruminant nutrition. Academic Press, New York.

Mitchell, R.B., K.J. Moore, L.E. Moser, J.O. Fritz, and D.D. Redfearn. 1997. Predicting developmental morphology in switchgrass and big bluestem. Agron. J. 89:827–832.

Mitchell, R.B., K.J. Moore, L.E. Moser, J.O. Fritz, and D.D. Redfearn. 2000. Predicting forage quality in switchgrass and big bluestem. Agron. J. (In review).

Moore, J.E. 1980. Forage crops. p. 61–91 *In* C. S. Hoveland (ed.) Crop, quality, storage, and utilization. ASA, CSSA, and SSSA, Madison, WI.

Moore, K.J., and J.H. Cherney. 1986. Digestion kinetics of sequentially extracted cell wall components of forages. Crop Sci. 26:1230–1234.

Moore, K.J., and R.D. Hatfield. 1994. Carbohydrates and forage quality. p. 229–280. *In* G.C. Fahey, Jr. (ed.) Forage quality, evaluation, and utilization. ASA, CSSA, and SSSA, Madison, WI.

Moore, K.J., L.E. Moser, K.P. Vogel, S.S. Waller, B.E. Johnson, and J.F. Pedersen. 1991. Describing and quantifying growth stages of perennial forage grasses. Agron. J. 83:1073–1077.

Moore, K.J., K.P. Vogel, A.A. Hopkins, J.F. Pedersen, and L.E. Moser. 1993. Improving the digestibility of warm-season perennial grasses. p. 447–448. *In* Proc. XVII Int. Grassl. Cong., Palmerston North, New Zealand. 8–21 Feb. 1993. New Zealand Grassland Assoc., Palmerston North.

Reid, R.L., G,A, Jung, and W.V. Thayne. 1988. Relationships between nutritive quality and fiber components of cool season and warm season forages: a retrospective study. J. Anim. Sci. 66:1275–1291.

Sarkanen, K.V., and C.H. Ludwig. 1971. Definition and nomenclature. p. 1–18. *In* K.V. Sarkanen and C. H. Ludwig (ed.) Lignins: Occurrence, formation, structure and reactions. John Wiley & Sons, New York.

Smith, L.W., H.K. Goering, and C.H. Gordon. 1972. Relationships of forage compositions with rates of cell wall digestion and indigestibility of cell walls. J. Dairy Sci. 55:1140–1147.

Twidwell, E.K., K.D. Johnson, J.H. Cherney, and J.J. Volenec. 1988. Forage quality and digestion kinetics of switchgrass herbage and morphological components. Crop Sci. 28:778–782.

Twidwell, E.K., K.D. Johnson, J.A. Patterson, J.H. Cherney, and C.E. Bracker. 1991. Degradation of switchgrass anatomical tissue by rumen microorganisms. Crop Sci. 30:1321–1328.

Van Soest, P.J. 1994. Nutritional ecology of the ruminant. Comstock Publ., Ithaca, NY.

Van Soest, P.J., J.B. Robertson, and B.A. Lewis. 1991. Methods for dietary fiber, neutral detergent fiber, and nonstarch polysaccharides in relation to animal nutrition. J. Dairy Sci. 74:3583–3597.

Vogel, K.P. F.A. Haskins, H.J. Gorz, B.A. Anderson, and J.K. Ward. 1991. Registration of 'Trailblazer' switchgrass. Crop Sci. 31:1388.

Vogel, K.P. A.A. Hopkins, K.J. Moore, K.D. Johnson, and I.T. Carlson. 1996. Registration of 'Shawnee' switchgrass. Crop Sci. 36:1713.

Wattiaux, M.A., L.D. Satter, and D.R. Mertens. 1992. Effect of microbial fermentation on functional specific gravity of small forage particles. J. Anim. Sci. 70:1262–1270.

Wedig, C.L., E.H. Jaster, and K.J. Moore. 1987. Hemicellulose monosaccharide composition and in vitro disappearance of orchardgrass and alfalfa hay. J. Agric. Food Chem. 35:214–218.

Wilkie, K.C.B. 1979. The hemicelluloses of grasses and cereals. Adv. Carbohydr. Chem. Biochem. 36:215–264.

Wilson, J.R. 1993. Organization of forage plant tissues. p. 1–32. *In* H.G. Jung et al. (ed.) Forage cell wall structure and digestibility. ASA, CSSA, and SSSA, Madison, WI.

Wilson, J.R., and P.M. Kennedy. 1996. Plant and animal constraints to voluntary feed intake associated with fibre characteristics and particle breakdown and passage in ruminants. Aust. J. Agric. Res. 47:199–225.

Wilson, J.R, and D.R. Mertens. 1995. Cell wall accessibility and cell structure limitations to microbial digestion of forage. Crop Sci. 35:251–259.

Windham, W.R., F.E. Barton, II, and D.S. Himmelsbach. 1983. High-pressure liquid chromatographic analysis of component sugars in neutral detergent fiber for representative warm- and cool-season grasses. J. Agric. Food Chem. 31:471–475.

3 Morphology of Germinating and Emerging Warm-Season Grass Seedlings

Lowell E. Moser

University of Nebraska
Lincoln, Nebraska

Establishment of warm-season grass seedlings is a slow and risky process. Since warm-season grass seed is generally expensive, land may have to be taken out of production for several seasons, and seeding failures are common, producers are often unwilling or unable to face the economic risk associated with establishing warm-season grasses. Therefore, producers often use cool-season grasses when they really need summer pasture because they achieve a usable stand of cool-season grasses much more quickly and reliably. Understanding morphological development during germination and seedling development process is essential to select appropriate seeding and seeding management practices. The weak seedling vigor often attributed warm-season grasses relates to physiological and morphological aspects of seedling development. With appropriate management practices, warm-season grass seedling vigor, per se, is not the factor limiting establishment.

A clear understanding of grass seedling morphology and the implications for establishment will enable managers to design and use management practices that minimize risk and allow seedlings to establish as perennial plants.

GERMINATION

The germination and pre-emergence phase of development is a heterotrophic process (McKell, 1972) where the energy for the growth process is provided by seed resources. Germination potential, including vigor, is affected by genetic factors and environmental conditions during seed development as well as the prevailing conditions at germination time. Seed dormancy may be prevalent in some species and cultivars (Coukos, 1944). In some instances, seed dormancy may be affected by slow morphological development of the embryo (Booth & Haferkamp, 1995) but in many warm-season grasses used in the Great Plains area the mechanism of dormancy is not well established. Generally, the dormancy of fresh seed can be broken with a 4-wk wet chill treatment (Emal & Conard, 1973), suggesting some sort of biochemical dormancy.

Copyright © 2000. Crop Science Society of America and American Society of Agronomy, 677 S. Segoe Rd., Madison, WI 53711, USA. *Native Warm-Season Grasses: Research Trends and Issues.* CSSA Special Publication no. 30.

The Embryo

A warm-season grass seed consists of an embryo, the endosperm, and a seedcoat. The endosperm is the heaviest component of most warm-season grass seeds and serves as the energy source for early root and shoot growth. Warm-season grass seed embryos often are relatively larger than those of cool-season grasses (Reeder, 1957). Seed size often affects the size of the embryo. Booth and Haferkamp (1995) concluded that larger seeds produce larger seedlings with a greater capacity to extract water from the soil thus producing greater photosynthetic area. A large pool of carbohydrate in the seed also provides more energy reserves to be used in the early growth process. Larger seedlings are able to compete more effectively for resources than small seedlings, which could influence their development and the establishment process.

The major components of the embryo are the plumule, radicle, scutellum (single cotyledon), coleoptile, and coleorhiza (Fig. 3–1). The plumule will develop into the shoot and the radicle into the primary root system of the plant. The area between the scutellar node (first node of the plant) and the coleoptilar node (second node of the plant) is called the subcoleoptile internode, or mesocotyl by some authors, and serves an important role in seedling emergence of warm-season grasses. Ries and Hofmann (1991) present standardized terminology to use in discussing seedling morphology. They use mesocotyl while many authors use subcoleoptile internode.

The embryo is imbedded in the endosperm and the scutellum is pressed against the endosperm. It appears that the main function of the scutellum in grass embryos is to absorb catabolites from the endosperm during germination (Brown, 1960).

The coleoptile is a tapered covering over the plumule and provides protection to the shoot as it penetrates the soil during emergence. The coleorhiza is a covering over the radicle but is penetrated when the primary root emerges from the seed. An upward protrusion called the epiblast is present on the coleorhiza, opposite the scutellum, in the Chlorideae and Zoysiaeae tribes. It is not present on grasses belonging to the Paniceae and Andropogoneae tribes (Reeder, 1957), which contain most of the warm-season grasses of interest in the Midwest and Great Plains.

Interpretation of the origin, purpose, and function of various embryo components has been debated. Brown (1960, 1965) has written well-documented reviews of the formation of the concepts of grass embryo morphology during the past 300 yr. Embryos may differ in the actual morphology of the structures they contain. Reeder (1957) illustrates the range of grass embryos from 300 species representing the various grass tribes and shows how embryo differences may be used in grass systematics.

Germination and Pre-Emergent Development

Germination for warm-season grasses begins with imbibition. After imbibition the coleorhiza and radicle penetrate the seedcoat and, in turn, the primary root penetrates the coleorhiza and grows into the soil. This becomes the first visible sign that the seed has germinated (Fig. 3–2). Shortly afterward the coleoptile, which encloses the shoot, penetrates the seed. With switchgrass (*Panicum virgatum* L.) and sand bluestem (*Andropogon gerardii* var. *paucipilus* (Nash) Fern.] Ohlenbusch (1975) reported that the expanding scutellar node ruptures the seedcoat and the coleorhiza and coleoptile emerge. Then the radicle penetrates the coleorhiza.

Normal germination occurs when both the root and shoot penetrate the seedcoat. Abnormal germination can occur where only the shoot or the root emerge from the seed. Tischler and Voigt (1981) reported that there was an increase in seedlings of weeping lovegrass [*Eragrostis curvula* (Schrad.) Nees.] and kleingrass (*Panicum coloratum* L.) with no primary root when they were grown at 35°C compared with seedlings grown at 20°C. They also reported that lighter (less mature) seed produced a higher incidence of seedlings with no primary root than heavier seed. In kleingrass, the exposure of the coleoptile to light led to normal primary root growth but this did not happen with weeping lovegrass. The pre-emergence seedling shoot and

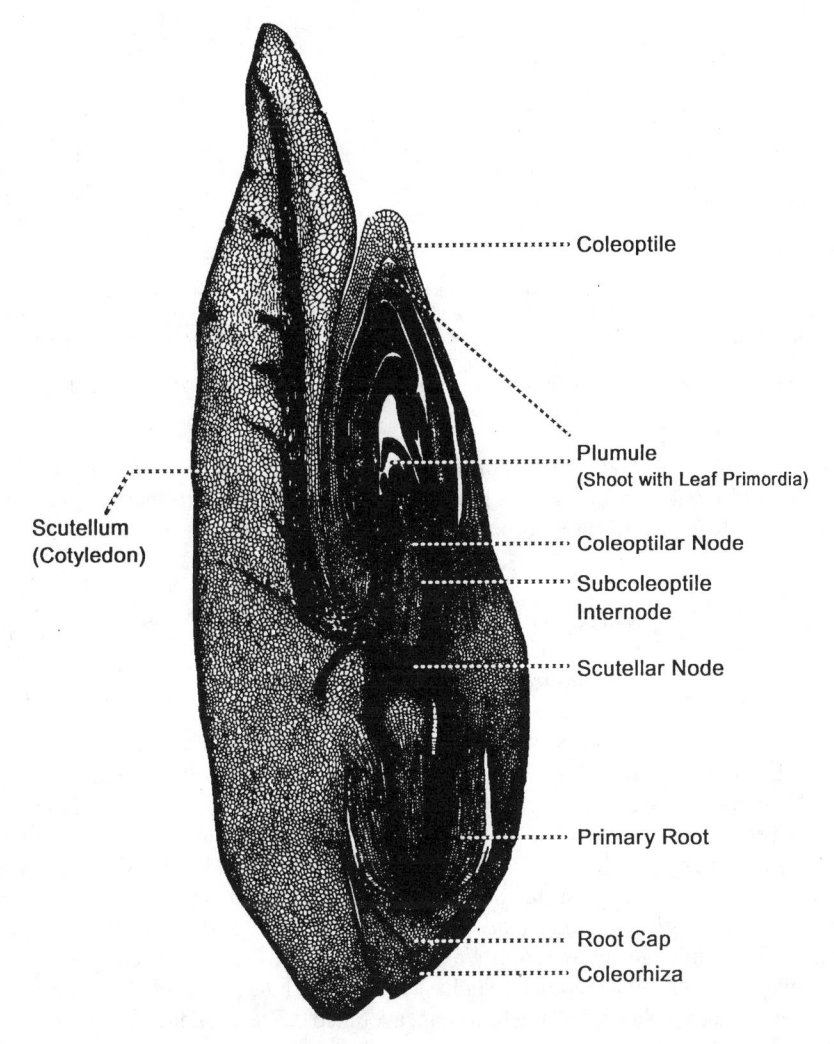

Fig. 3–1. The grass embryo (Avery, 1930).

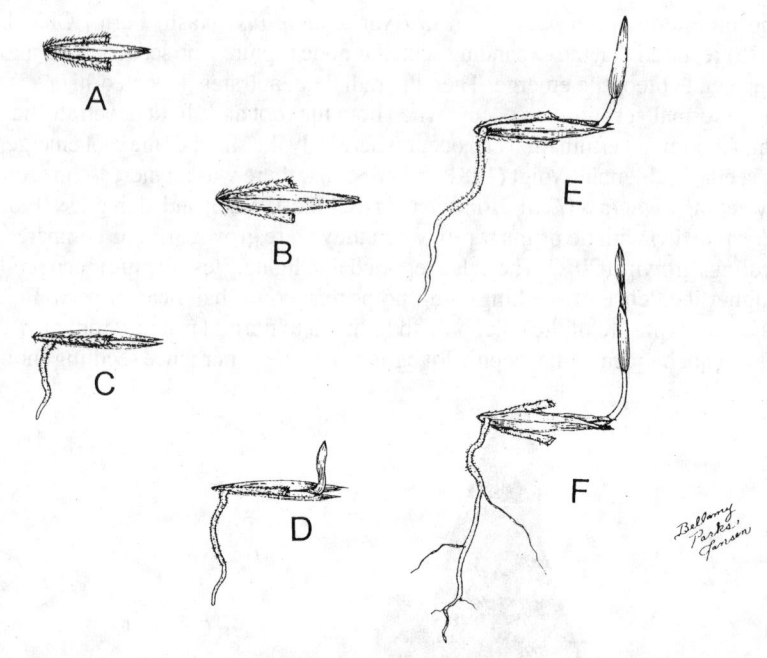

Fig. 3–2. Morphological development of big bluestem from germination until seedling emergence. A: Big bluestem seed. B: Imbibed seed before visible signs of germination. C: Radicle penetration of the seed and formation of primary root. D: Continued elongation of primary root and emergence of coleoptile from seed. E: Continued development of primary root and beginning of elongation of sub-coleoptile internode. F: Emerging seedling; branching of primary root and emergence of shoot from the coleoptile. Drawings by Bellamy Parks Jansen.

root development process of big bluestem (*Andropogon gerardii* var. *gerardii* Vitman) is illustrated in Fig. 3–2A–F. Grasses have hypogeal germination since the scutellum (cotyledon) remains below the surface of the soil.

MORPHOLOGY OF EMERGING SEEDLINGS

Primary and Seminal Lateral Roots

Seminal lateral roots (sometimes called seminal roots) may develop at the scutellar node on some grasses (Hyder, 1974). Hoshikawa (1969) reported that seminal lateral roots do not form on warm-season grasses. Newman and Moser (1988a), however, found a few seminal lateral roots on warm-season grass seedlings at the three-leaf emergence stage. An average of 0.7, 0.5, 0.4, and 0.4 seminal lateral roots were present on switchgrass, indiangrass [*Sorghastrum nutans* (L.) Nash], caucasian bluestem [*Bothriochloa caucasica* (Trin.) C.E. Hubb.], and sand bluestem, respectively, while blue grama [*Bouteloua gracilis* (H.B.K.) Lag. ex Steud.], sideoats grama [*Bouteloua curtipendula* (Michx.) Torr.], little bluestem [*Schizachrium scoparium* (Michx.) Nash], and big bluestem, averaged 0.3 seminal lateral roots or less per seedling.

MORPHOLOGY OF GRASS SEEDINGS

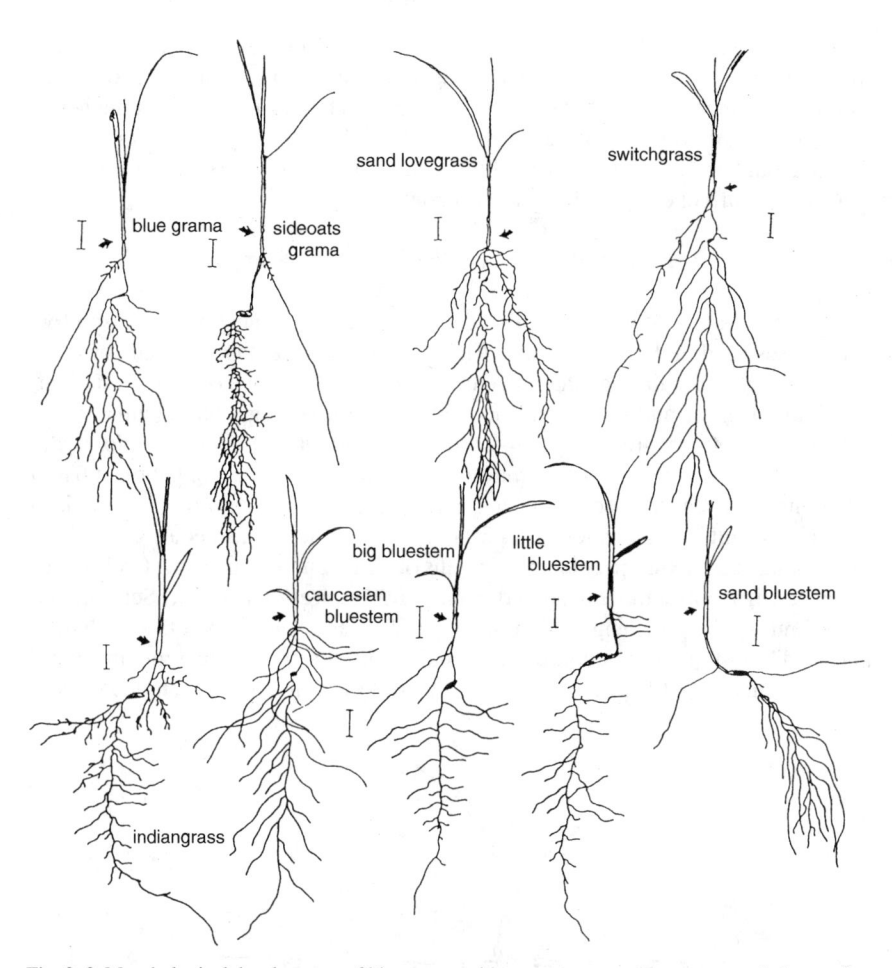

Fig. 3–3. Morphological development of blue grama, sideoats grama, sand lovegrass, switchgrass, indiangrass, caucasian bluestem, big bluestem, little bluestem, and sand bluestem at the three leaf emergence stage. Arrow points to the soil level. Vertical bars represent 1 cm. (Newman, 1986).

Warm-season grasses must rely on the primary root system for water absorption early in the life of the seedling. The primary root systems of various grasses differ in morphology (Fig. 3–3)(Newman, 1986), which may affect water and nutrient uptake. Tischler and Monk (1980) reported that the terminal pre-emergence length of the primary root (i.e., primary root growth stopped) of kleingrass was 10 mm and ranged from 0 to 30 mm until the coleoptile reached the soil surface, at which time primary root growth resumed. Resumption of growth did not depend on photosynthate since 5 min of light per day promoted growth as well as continuous light.

There is a great deal of morphological diversity among seedling grass root systems. Hoshikawa (1969) classified 219 species of 88 genera and found that seedling root morphology differed among tribes of grasses. Generally, species of the same genus and, even of the same tribe, had similar root morphology. With switchgrass and sand bluestem, Ohlenbusch (1975) found three types of seminal root systems. One type of plant had a long single seminal root and a second type had freely branching roots but did not root deeply. The third was a combination of both types that resulted in the heaviest root system. The three types of root systems occurred randomly, so this seminal rooting pattern appeared to be under genetic control and not influenced greatly by environment.

Coleoptile and Subcoleoptile Internode Elongation

Panicoid morphology (Hoshikawa, 1969) describes most warm-season forage grasses used in the Great Plains. The coleoptile covers the shoot and protects it during emergence through the soil. Once the coleoptile reaches light (soil surface) the shoot emerges and leaves expand (Fig. 3–4). The elongation potential of the coleoptile is not very great on warm-season (panicoid) grasses. For example, Hyder et al. (1971) reported that blue grama had a potential coleoptile elongation of 6 mm while the cool-season species, crested wheatgrass [*Agropyron desertorum* (Fisch ex Link) Schult.], had a potential of 30 mm. Little bluestem, sideoats grama, and blue grama had maximum coleoptile lengths of 7, 11, and 9 mm, respectively (Redmann & Qi, 1992). Prairie sandreed [*Calamovilfa longifolia* (Hook) Scribn.] had a maximum coleoptile length of 8 mm (Maun & Riach, 1981). In both switchgrass and sand bluestem the coleoptile grew rapidly until it reached 6 to 8 mm in length. Then the subcoleoptile internode elongated until the coleoptile penetrated the soil

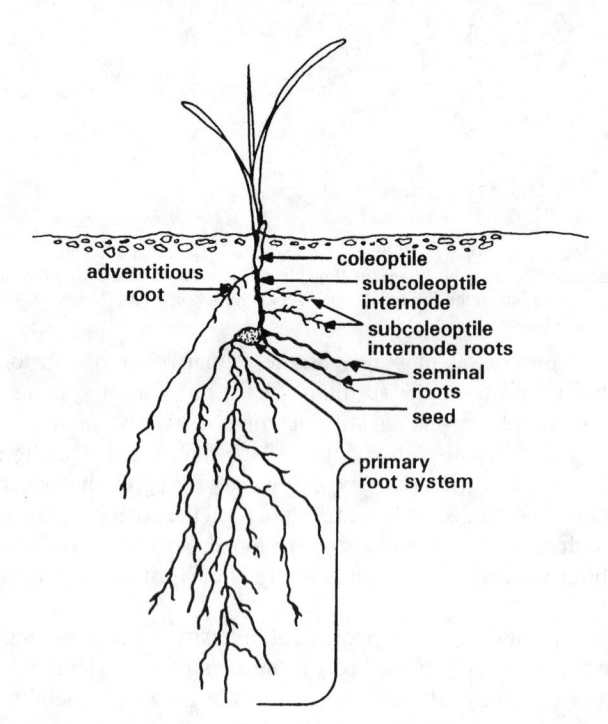

Fig. 3–4. The generalized grass seedling (Newman & Moser, 1988a).

or the first leaf exerted through the coleoptile tip. The role of the first leaf appears to serve as a transition from the seed to the shoot supplying energy for growth (Ohlenbusch, 1975).

Since protection of the fragile shoot is essential during emergence, the coleoptile of most warm-season grasses seems to have the potential to protect shoots from only very shallowly planted grasses. The subcoleoptile internode (between the scutellar and coleoptilar node) (Fig. 3–1 to 3–4) elongates when no light signal is received by the coleoptile, pushing up the coleoptilar node. Thus, the shoot with the protective coleoptilar shield is able to reach the soil surface undamaged.

Once the coleoptile receives a light signal (specifically red light, around 660 nm wavelength), subcoleoptile internode elongation ceases (van Overbeek, 1936). Subcoleoptile internode ceases because red light reduces the auxin supply from the coleoptile. Exogenously applied auxin allowed elongation of the subcoleoptile internode to continue even though the coleoptile was in the light (590 to 730 nm, 630 nm maximum)(Vanderhoef & Briggs, 1978). Tischler and Voigt (1993) reported that in some cases there can be excessive subcoleoptile internode elongation and the coleoptilar node can end up above the soil surface in sideoats grama, blue grama, and kleingrass. This probably occurs when there is excessive competition for red light with existing vegetation, resulting in light at the soil surface being enriched in far-red relative to the red wavelengths. Also, when seedlings develop rapidly and emerge early in the night, the red light signal is not received until morning, after the coleoptilar node is above the soil surface. They reported genetic variation regarding that trait so progress could be made in breeding cultivars without excessive subcoleoptile internode elongation.

In contrast, most cool-season grasses have long coleoptiles that are able to reach the soil surface and the red light signal without subcoleoptile internode elongation. Many cool-season grasses, however, have a small amount of subcoleoptile internode elongation (Hoshikawa, 1969; Newman & Moser, 1988a). The ability of warm-season grasses to elongate the subcoleoptile internode to a considerable extent may give them an advantage over cool-season grasses when planted rather deeply. Although emergence was only about 10%, Maun and Riach (1981) documented that prairie sandreed had a 8-mm coleoptile and a 70-mm subcoleoptile internode when it emerged at an 8-cm planting depth. There is generally more potential for subcoleoptile internode elongation in warm-season grasses than for coleoptile elongation in cool-season grasses (Tischler & Voigt, 1987). Western wheatgrass [*Pascopyrum smithii* (Rydb.) A. Love.] , however, had a higher value for the length of the subcoleoptile internode plus the coleoptile length than did blue grama (Ries & Hofmann, 1995).

Boyd and Avery (1936) stated that the anatomy of the subcoleoptile internode was unique, anatomically, being neither root-like or stem-like. Brown (1960) refers to it as a structure unique to the embryo. Redmann and Qi (1992), however, indicated that the vascular anatomy of the subcoleoptile internode of blue grama, sideoats grama, and little bluestem was like the roots while that of three wheatgrasses was like that of the stems.

Hoshikawa (1969) reported that most warm-season grasses develop roots on the subcoleoptile internode (Fig. 3–3) and these roots may be quite numerous and long. They are probably important in water absorption early in the life of the seedling because these grasses develop few or no seminal lateral roots. The subcoleoptile internode roots may have special value when the primary root is damaged or missing (Tischler & Voigt, 1987). Newman and Moser (1988a) found subcoleoptile internode roots on 90 to 100% of sand lovegrass [*Eragrostis trichodes* (Nutt.) Wood], big bluestem, caucasian bluestem, indiangrass, and switchgrass seedlings, on 87% of little bluestem seedlings, on 71% of sand bluestem seedlings, and only on 4 to 7% of the seedlings of blue and sideoats grama when the seedlings were at the three-leaf emergence stage. Most cool-season grasses did not have subcoleoptile internode roots at the three-leaf stage.

The short coleoptile and resulting elongation of the subcoleoptile internode in warm-season grasses places the coleoptilar node or the seedling crown near the soil surface while it remains near the planting depth in cool-season grasses. The meristematic tissue is in the upper part of the subcoleoptile internode producing new cells at the top of the internode compared with being at the lower part of the other stem internodes producing cells at the bottom (Brown, 1960). This makes the subcoleoptile internode different and especially effective in pushing up the coleoptilar node without bending (Hyder et al., 1971). Regardless of the depth of planting of warm-season grasses, within in the inherent level for the seedling to emerge, the seedling crown is placed at the same point near the soil surface.

MORPHOLOGICAL DEVELOPMENT OF EMERGED SEEDLINGS

Root Development

The coleoptilar node is the first point of origin of adventitious roots that is the key to seedling establishment. Simanton and Jordan (1986) found that rapid root elongation or relatively high root/shoot ratio, per se, could not be related to success or failure of seedling establishment of sideoats grama, Cochise lovegrass (*Eragrostis lehmanniana* Nees. × *Eragrostis trichophora* Cass and Dur.), or blue panic (*Panicum antidotale* Retz.).

Primary Root Longevity

It often has been reported that the primary root and other seminal roots are short-lived so adventitious (permanent) roots must be initiated before the seedling can become established. Often blue grama seedlings in the field die about 6 to 10 wk after emergence if adventitious roots do not form (Briske & Wilson, 1980). Ries and Svejcar (1991) stated that blue grama seedlings must have at least two adventitious roots rooted to a 10-cm depth before they considered a seedling established. Seedling death in blue grama, however, is not caused by some inherent longevity of the seminal root system since blue grama seedlings in the greenhouse grew actively and continued tiller recruitment for 22 wk when restricted to seminal roots (van der Sluijs & Hyder, 1974). Likewise, when native grasses were restricted to seminal roots they remained functional, were deep and widespread, and absorbed water for 3.5 to 4 mo (Weaver & Zink, 1945). In the van der Sluijs and Hyder (1974) study, where seedlings were restricted to primary roots, tillering was initiated at 3 wk and continued at a rate of 0.165 tillers per day. Leaf length on primary shoots reached 80 cm at 6 to 7 wk, and older leaves died afterward. Maximum total green leaf length of tillers reached 250 to 350 cm at 13 to 14 wk.

Water Transport

Wilson et al. (1976) attributed the water transport capacity of the subcoleoptile internode as limiting further leaf growth. Before adventitious root initiation all of the water absorbed by the primary root system and used by the tops must pass through the subcoleoptile internode. Seedlings can only expand leaf area to the maximum that can be supported by water absorption and transport by the seminal root system and subcoleoptile internode. In blue grama the subcoleoptile internode was very fine (135 µm stele diam) and the average radius of xylem vessels was only 5 µ (Wilson et al., 1976). The radius is similar to that for blue grama, sideoats grama, and little bluestem reported by Redmann and Qi (1992). Wilson et al. (1976) reported a total xylem cross section of 580 μ^2, which was about 4% of the cross sectional area of the subcoleoptilar internode. This compares with about 3000 μ^2 of xylem cross sectional area in an adventitious root, about five times more than that of the subcoleoptile internode. Not all subcoleoptile internodes may be as thin as that of blue grama (Fig. 3–3), but this probably is the main reason seedlings cannot survive until adventitious roots are formed. Redmann and Qi (1992) found that blue and sideoats grama had smaller effective xylem vessel radii in their subcoleoptile internodes compared with little bluestem. With increasing depth of planting there generally was a reduction in number of xylem vessels in the subcoleoptile internode. Vessel diameter remained relatively constant with planting depth. Since hydraulic conductance is affected by both vessel diameter and path length, it was less in warm-season grass seedlings (with longer subcoleoptile internodes) than in cool-season grasses and decreased markedly with depth of planting.

Adventitious Root Initiation

Since the coleoptilar node, or seedling crown, of warm-season grasses is placed very near the soil surface they are at a disadvantage in initiation of adventitious roots. Adventitious root development may be delayed or not occur under dry conditions. This may be why blue grama seedlings often do not survive past 6 to 10 wk in the field. Wilson and Briske (1979) reported that not only must the soil remain moist from 2 to 4 d for blue grama seeds to germinate, but in addition, the soil surface must remain moist for 2 to 4 d sometime 2 to 8 wk after emergence in order for adventitious roots to form. Adventitious roots grew out of tillering crowns of blue grama only when damp, cloudy weather persisted for 2 to 3 d (van der Sluijs & Hyder, 1974). Newman and Moser (1988b) illustrated the same principle with big bluestem, indiangrass, and switchgrass. In 1 yr under dry conditions there were very few adventitious roots until after a 4-d rainfall period after which they increased rapidly. In the second year when rainfall was distributed more uniformly adventitious root development occurred continuously.

Depth of planting had no effect on the number of adventitious roots developed on warm-season grasses since the coleoptilar node is placed near the soil surface regardless of the planting depth (Newman & Moser, 1988b). Ries and Hofmann (1995) found that adventitious root numbers decreased significantly when planting depth exceeded 51 mm for sideoats grama. Since subcoleoptile internode elongation is small with cool-season grasses, the coleoptilar node stays near the planting depth where water relations are much better for adventitious root development.

There is a juvenile period where a seedling cannot initiate adventitious roots under favorable environmental conditions. In kleingrass, seedling age is more important that seedling mass in influencing adventitious root initiation, so selecting for the largest seedling may not necessarily shorten the interval between planting and initiation of adventitious roots (Tischler et al., 1989). Wilson and Briske (1979) found that blue grama could initiate roots as early as 2 wk after emergence under proper moisture conditions. Newman and Moser (1988a) found very few adventitious roots on grasses of the Andropogoneae tribe at the three-leaf emergence stage but most switchgrass plants had one adventitious root at that stage. At 3 wk of age greenhouse grown big bluestem, little bluestem, sideoats grama, switchgrass, and indiangrass seedlings had 0 to 1, 1 to 2, 2 to 3, 2 to 3, and 2 to 4 adventitious roots, respectively, but buffalograss *[Buchloe dactyloides* (Nutt.) Engelm.] had none (Weaver & Zink, 1945).

Growth of an emerged seedling is affected by a complex set of environmental factors (Fig. 3–5). Air temperature affects photosynthesis, other physiological processes, and growth of the shoot. Soil temperature affects cell division and growth at coleoptilar node, affecting vegetative growth and adventitious root development. Soil temperature and soil moisture potential at the root tips affect root growth (Ohlenbusch, 1975).

Temperature and Soil Water Relationships with Grass Seedling Growth Processes

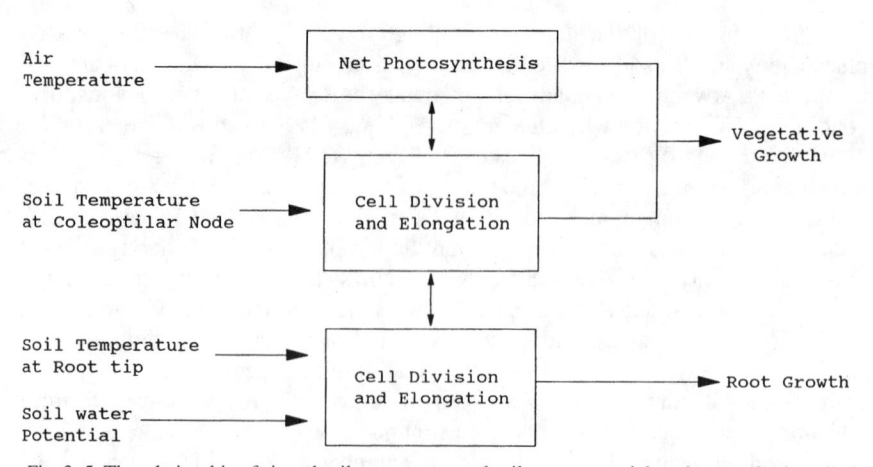

Fig. 3–5. The relationship of air and soil temperature and soil water potential to photosynthesis, cell division and elongation, and plant growth (Ohlenbusch, 1975).

QUANTIFYING GRASS SEEDLING MORPHOLOGY

Evaluation of the establishment status of warm-season grasses is important when making seeding management decisions on a field basis. The key factor in determining establishment of warm-season grasses is development of adventitious roots (Newman & Moser, 1988a; Ries & Svejcar, 1991). Although the relative shoot and root development is rather predictable within species, there is a differential in shoot and root development among grasses (Newman & Moser, 1988a). Therefore, roots and shoots need to be classified separately. A system to quantify grass seedling development (Moser et al., 1993) can be used to obtain the average developmental status of a grass seeding (Table 3–1). The average development of a seedling population can be calculated as the mean stage count for shoots (MSCS) or roots (MSCR). The mean stage count is simply the weighted mean of the indices assigned to the root or shoot stages and is calculated as:

MSCS = [(shoot index of seedling 1) + (shoot index of seedling 2)

+ ... + (shoot index of seedling n)]/(total number of seedlings)

MSCR = [(root index of seedling 1) + (root index of seedling 2)

+ ... + (root index of seedling n)]/(total number of seedlings)

When adventitious root development is well underway for a stand (e.g., MSCR = 5), the stand is rather well established and will be able to withstand considerable stress. Excavating and quantifying the number of seedlings necessary to

Table 3–1. Growth stages, numerical indices, and descriptions for staging grass seedling shoots and roots.

Stage	Index	Description
Shoot		
S1	1.0	Coleoptile emerged from seed
S2	2.0	Emergence of leaf from coleoptile
S3	3.0	Appearance of collar–ligule of first leaf
S4	4.0	Appearance of collar–ligule of second or third leaf†
S5	5.0	Appearance of collar–ligule of fourth to sixth leaf†
S6	6.0	Appearance of first secondary tiller
S7	7.0	Appearance of second secondary tiller
S8	8.0	Appearance of third secondary tiller
Root		
R1	1.0	Radicle emergence from seed
R2	2.0	Primary root branching–seminal root formation
R3	3.0	Appearance of first adventitious root
R4	4.0	Appearance of two to three adventitious roots†
R5	5.0	Appearance of four to six adventitious roots†
R6	6.0	Appearance of seven to ten adventitious roots†
R7	7.0	Appearance of ten or more adventitious roots†
R8	8.0	Appearance of first true rhizome

† The number of leaves or roots in these stages can be changed to provide flexibility that will accommodate a wide range of developmental variation among species.

determine the MSCR, however, would be time consuming. Relationships between MSCS, which is determined easily, and MSCR could be developed to predict MSCR. Decisions could be made concerning establishment status based on the predicted status of root development. The MSCS values of 5.0, and 5.1 are associated with the MSCR of 4 for switchgrass and big bluestem, respectively. Most big bluestem seedlings are at the four to six-leaf stage while switchgrass seedlings often have fewer leaves but have one secondary tiller at this time.

Seeding and seedling management practices should be related to morphological status of seedlings just as harvest management is related to grass morphology. An early seeding date will produce seedlings capable of initiating adventitious roots during late spring when rainfall is much more likely than in midsummer when late-planted warm-season grasses may be able to initiate adventitious roots. Determining the establishment status may give a producer the necessary insight to evaluate competitive ability and the need for clipping, herbicide treatment, or reseeding. Evaluating the population would be an accurate way to determine the morphological status, which could be a basis used for making management decisions.

REFERENCES

Avery, G.S. 1930. Comparative anatomy and morphology of embryos and seedlings of maize, oats, and wheat. Bot. Gaz. 89:1–39.

Booth, D.T., and M.R. Haferkamp. 1995. Morphology and seedling establishment. p. 239–290. *In* D.J. Bedunah and R.E. Sosebee (ed.). Wildland plants: Physiological ecology and developmental morphology. Soc. for Range Manage, Denver, CO.

Boyd, L., and G.S. Avery. 1936. Grass seedling anatomy: The first internode of *Avena* and *Triticum*. Bot. Gaz. 97:765–799.

Briske, D.D., and A.M. Wilson. 1980. Drought effects on adventitious root development in blue grama seedlings. J. Range Manage. 33:323–327.

Brown, W.V. 1960. The morphology of the grass embryo. Phytomorphology 10:215–223.

Brown, W.V., 1965. The grass embryo - a rebuttal. Phytomorphology 15:274–284.

Coukos, C.J. 1944. Seed dormancy and germination in some native grasses. J. Am. Soc. Agron. 36:337–345.

Emal, J.G., and E.C. Conard. 1973. Seed dormancy and germination in indiangrass as affected by light, chilling, and certain chemical treatments. Agron. J. 65:383–385.

Hoshikawa, K. 1969. Underground organs of the seedlings and the systematics of Gramineae. Bot. Gaz. 130:192–203.

Hyder, D.N., A.C. Everson, and R.E. Bement. 1971. Seedling morphology and seeding failures with blue grama. J. Range Manage. 24:287–292.

Hyder, D.N. 1974. Morphogenesis and management of perennial grasses in the United States. p. 89–98. *In* K.W. Kreitlow and R.H. Hart (ed.) Plant morphogenesis as the basis for scientific management of range resources. Proc. Workshop of the United States-Australia Rangelands Panel. USDA Misc. Publ. no. 1271. USDA, Washington, DC.

Maun, M.A., and S. Riach. 1981. Morphology of caryopses, seedlings, and seedling emergence of the grass *Calamovilfa longifolia* from various depths in sand. Oecologia 49:137–142.

McKell, C.M. 1972. Seedling vigor and seedling establishment, p. 74–89. *In* V.B. Youngner and C.M. McKell (ed.) The biology and utilization of grasses. Academic Press, New York.

Moser, L.E., K.J. Moore, M.S. Miller, S.S. Waller, K.P. Vogel, J.R. Hendrickson, and L.A. Maddux. 1993. A quantitative system for describing the developmental morphology of grass seedling populations. p. 317–318. *In* M.J. Baker et al. (ed.) Proc. 17th Int. Grassl. Cong., Palmerston North, New Zealand. 8–21 Feb. Keeling & Mundy Printing, Palmerston-North, New Zealand.

Newman, P.R. 1986. Seedling root development of forage grasses. M.S. thesis. Univ. of Nebraska, Lincoln.

Newman, P.R., and L.E. Moser. 1988a. Seedling root development and morphology of cool-season and warm-season forage grasses. Crop Sci. 28:148–151.

Newman, P.R., and L.E. Moser. 1988b. Grass seedling emergence, morphology, and establishment as affected by planting depth. Agron. J. 80:383–387.

Ohlenbusch, P.D. 1975. Emergence and early growth of two warm-season grasses. Ph.D. diss. Texas A&M Univ., College Station.

Redmann, R.E., and M.Q. Qi. 1992. Impacts of seeding depth on emergence and seedling structure in eight perennial grasses. Can. J. Bot. 70:133–139.

Reeder, J.H. 1957. The embryo in grass systematics. Am. J. Bot. 44:756–768.

Ries, R.E., and L. Hofmann. 1991. Research observations: Standardized terminology for structures resulting in emergence and crown placement of 3 perennial grasses. J. Range Manage. 44:404–407.

Ries, R.E., and L. Hofmann. 1995. Grass seedling morphology when planted at different depths. J. Range Manage. 48:216–223.

Ries, R.E., and T.J. Svejcar. 1991. The grass seedling: When is it established? J. Range Manage. 44:574–576.

Simanton, J.R., and G.L. Jordan. 1986. Early root and shoot elongation of selected warm-season perennial grasses. J. Range Manage. 39:63–67.

Tischler, C.R., and P.W. Monk 1980. Variability in root system characteristics of kleingrass seedlings. Crop Sci. 20:384–386.

Tischler, C.R., and P.W. Voigt. 1981. Non-genetic factors affecting primary root absence in lovegrass and kleingrass. Crop Sci. 21:427–430.

Tischler, C.R., and P.W. Voigt. 1987. Seedling morphology and anatomy of rangeland plant species. p. 5–13. In G.W. Frasier and R.A. Evans (ed.) Proc. of a Symp. on Seed and Seedbed Ecology of Rangeland Plants, Tuscon, AZ. 21–23 Apr. 1987. USDA Natl. Tech. Inf. Serv., Springfield, VA.

Tischler, C.R., and P.W. Voigt. 1993. Characterization of crown node elevation in Panicoid grasses. J. Range Manage. 46:436–439.

Tischler, C.R., P.W. Voigt, and E.C. Holt. 1989. Adventitious root initiation in kleingrass in relation to seedling size and age. Crop Sci. 29:180–183.

Vanderhoef, L.N., and W.R. Briggs. 1978. Red light-inhibited mesocotyl elongation in maize seedlings. Plant Physiol. 61:534–537.

van Overbeek, J. 1936. Growth hormone and mesocotyl growth. Rec. Trav. Bot. Neerl. 33:333–340.

Van der Sluijs, D.H., and D.N. Hyder. 1974. Growth and longevity of blue grama seedlings restricted to seminal roots. J. Range Manage. 27:117–119.

Weaver, J.E., and E. Zink. 1945. Extent and longevity of the seminal roots of certain grasses. Plant Physiol. 20:359–379.

Wilson, A.M., and D.D. Briske. 1979. Seminal and adventitious root growth of blue grama seedlings on the Central Plains. J. Range Manage. 32:209–213.

Wilson, A.M., D.N. Hyder, and D.D. Briske. 1976. Drought resistance characteristics of blue grama seedlings. Agron. J. 68:479–484.

4 Developmental Morphology and Tiller Dynamics of Warm-Season Grass Swards

Rob B. Mitchell
Texas Tech University
Lubbock, Texas

Lowell E. Moser
University of Nebraska
Lincoln, Nebraska

Growth and developmental morphology are important considerations in warm-season grass management. Although growth and developmental morphology are positively correlated, the two processes are not synonymous (Frank, 1996). Growth refers to the increase in dry weight resulting from the expansion of leaves, stems, and reproductive structures (Frank, 1996). Developmental morphology refers to the predictable series of changes in structure and arrangement of plant components associated with plant maturity (Esau, 1960). Changes in the developmental morphology of grasses affects decision making such as initial grazing, harvest for hay, and grazing of regrowth, or it may provide evidence of a management problem. Additionally, management practices such as herbicide application, fertilizer application, and prescribed burning may be applied based on the developmental morphology of the target species. Due to the exhibition of distinct contrasts to developmental morphology and the prominence and extensive research conducted on switchgrass (*Panicum virgatum* L.) and big bluestem (*Andropogon gerardii* Vitman), these species will be the focus of this chapter. Much of the research conducted on switchgrass and big bluestem is applicable to other warm-season grass species, especially those that are determinate in growth habit. Each species, however, has a particular pattern of developmental morphology and tiller recruitment.

DEVELOPMENTAL MORPHOLOGY

In general, developmental morphology is similar among grass species with only minor variations separating growth forms (Briske, 1991). Temperature and photoperiod are important factors controlling the rate of developmental morphology in warm-season grasses (Briske, 1991; Gillen & Ewing, 1992). Air temperature is the main environmental factor determining the rate of developmental morphology

---

Copyright © 2000. Crop Science Society of America and American Society of Agronomy, 677 S. Segoe Rd., Madison, WI 53711, USA. *Native Warm-Season Grasses: Research Trends and Issues.* CSSA Special Publication no. 30.

(Frank, 1996). Developmental morphology within a species has been reported to have strong linear relationships to accumulated growing degree days (GDD) and day of the year (Kalu & Fick, 1981; Hendrickson, 1992; Mitchell et al., 1997a). The relationship between developmental morphology and day of the year can be partially attributed to the process of floral induction that occurs in response to photoperiodic stimulus (Briske, 1991). Many plant species have photoperiod requirements for floral induction (Salisbury & Ross, 1985). Switchgrass and big bluestem are photoperiod sensitive and require shortening day length for floral induction, whereas blue grama [*Bouteloua gracilis* (H.B.K.) Lag. *ex* Steud.] is a plant of indeterminate day length response, and will flower throughout the summer (Benedict, 1941). Indiangrass [*Sorghastrum nutans* (L.) Nash] appears to be intermediate in its flowering behavior, displaying complete or partial inhibition of flowering when days are too short or too long (Allard & Evans, 1941). Most vegetative growth of switchgrass, big bluestem, and indiangrass terminates with inflorescence development and, therefore, is determinate in growth habit (Dahl & Hyder, 1977). Following floral induction, grass tillers advance to the seed ripening stages, growth stops, and tiller senescence occurs.

Canopy Development

Canopy architecture influences many plant canopy processes and must be considered when describing the interaction between plants and the environment (Welles & Norman, 1991). Canopy architecture affects forage plant physiology, quality of forage harvested by grazing animals, and animal grazing patterns (Nelson & Moser, 1994). Architecture of the grass sward is continually changing and is a function of the tiller morphology at various growth stages and the growth stage distribution within the tiller population (Moore & Moser, 1995). Quantification of the developmental morphology of tiller populations provides information for understanding these architectural changes in the grass sward. Additionally, canopy architectural measurements such as leaf area index (LAI) can be related to relative light interception, forage productivity, forage quality, forage availability, and forage accessibility to grazing livestock.

The phytomer is the basic modular unit of growth in grass tillers and consists of a leaf blade, leaf sheath, node, internode, and axillary bud (Hyder, 1972; Briske, 1991; Fig. 4–1). Phytomers originate from leaf primordia on the apical meristem (Fig. 4–2). A series of phytomers form the grass tiller, which consists of a single apical meristem, a stem, leaves, roots, nodes, dormant buds, and if reproductive, a potential inflorescence (Hyder, 1972; Vallentine, 1990; Fig. 4–1). Tillers form from buds located in the leaf axils of the lower internodes of the primary stem or other tillers. Grass tillers that arise from an individual tiller, crown, rhizome, or stolon and are of the same genotype form the grass plant (Moore & Moser, 1995). Grass plants collectively form a sward.

A grass leaf is composed of a leaf sheath and leaf blade. New leaves are generated by cell division and pushed upward by expansion at the basal meristem that results in the linear aspect of the entire leaf (Mauseth, 1988). Leaf blades emerge through the whorl and extend to the top of the canopy in grass canopies (Allard et al., 1991). The oldest leaves of a grass tiller have the lowest level of insertion from

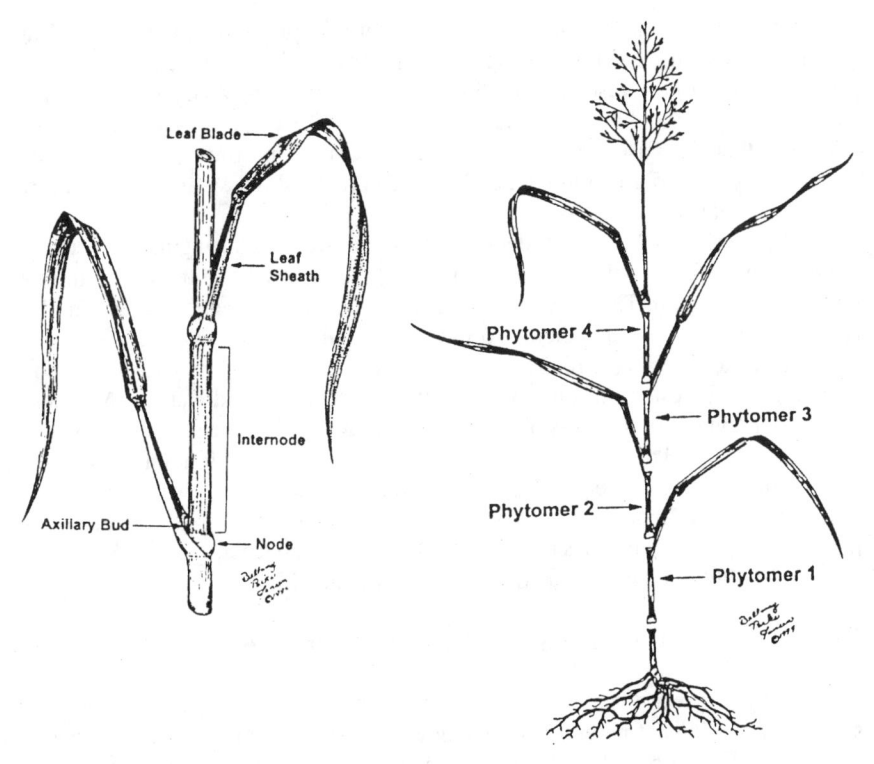

DEVELOPMENTAL MORPHOLOGY AND TILLER DYNAMICS

Fig. 4–1. Illustration of (a) the grass phytomer which consists of a leaf blade and sheath, the internode, the node, and the axillary bud located below the point of sheath attachment, and (b) a grass tiller showing the arrangement of phytomers (Moore & Moser, 1995).

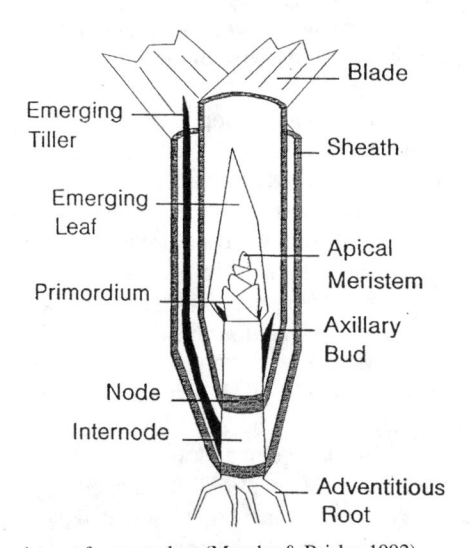

Fig. 4–2. The apical meristem of a grass plant (Murphy & Briske, 1992).

the plant base, whereas new leaves have a higher insertion level on the plant (Wilson, 1976; Walton, 1983). Leaf length in grass species is influenced by the transport limitations of the vascular bundles (Mauseth, 1988). In green panic (*Panicum maximum* var. trichoglume), leaf length and area increased progressively up to the tenth leaf, then decreased to the flag leaf. Leaves of high insertion levels developed more slowly, stayed green longer, and senesced more slowly than those of a low insertion level (Wilson, 1976).

Progression of leaf and node development in warm-season grasses occurs in a predictable manner for a species from one growing season to the next within a geographical region. The maximum number of collared leaves (V_{max}) and nodes (E_{max}) present prior to advancing to the elongation and reproductive stages, respectively, was determined for nondefoliated tillers of 'Trailblazer' switchgrass and 'Pawnee' big bluestem in four environments in Nebraska and Kansas (Mitchell, 1995). For switchgrass, the average V_{max} was 3 and the average E_{max} was 6. The V_{max} ranged from 3 to 4 and the E_{max} ranged from 6 to 7 in the four environments. For big bluestem, the average V_{max} was 5 and the average E_{max} was 6. The V_{max} ranged from 4 to 5 and the E_{max} ranged from 4 to 7 in the four environments. Similarly, Redfearn et al. (1997) determined the V_{max} was 4 and the E_{max} was 7 for 'Cave-in-Rock' switchgrass grown at Ames, IA, and Mead, NE.

Quantifying Developmental Morphology

The grass sward is a population of individual plants. Even in single-species stands, populations of perennial forage grasses represent numerous related genotypes rather than a pure line due to the open-pollinated nature of most perennial grasses (Moore & Moser, 1995). Management practices are applied at the sward or population level, so methods that quantify developmental morphology are needed that account for the inherent variability in developmental morphology of perennial grasses (Moore & Moser, 1995). Attempts to visually quantify the morphology of grass swards will probably overestimate actual morphology since sward maturity typically describes the most visually dominant tillers in the population (Mitchell et al., 1997b).

Numerous systems have been developed to accurately quantify the growth and development of plants (Vanderlip, 1972; Haun, 1973; Zadoks et al., 1974; Fehr & Caviness, 1977; Kalu & Fick, 1981; Simon & Park, 1983; Moore et al., 1991; Sanderson, 1992). Many of these systems were intraspecific or difficult to apply in the field. Kalu and Fick (1981) presented a morphological staging system for alfalfa (*Medicago sativa* L.) that included 10 growth stages ranging from early vegetative to ripe seed pod. This system included two methods for quantifying morphological stage of alfalfa shoot populations based on mean stage by count (MSC) and mean stage by weight (MSW).

In grasses, Phillips et al. (1954) used six general stages ranging from vegetative stage to seeds at dough stage. The vegetative stage included a broad spectrum of tillers that ranged from early elongation to the boot stage and lacked identification of nonelongated tillers. Haun (1973) developed a system for quantifying wheat (*Triticum aestivum* L.) development that integrated the number of leaves developed and the rate of development of the next older plant part into plant devel-

Table 4–1. Primary and secondary growth stages, numerical indices, and descriptions for staging development of perennial grasses (Moore et al., 1991).

Stage	Index	Description
Vegetative: Leaf development		
V0	1.0	Emergence of first leaf
V1	(1/N†) + 0.9	First leaf collared
V2	(2/N) + 0.9	Second leaf collared
Vn	(n/N) + 0.9	Nth leaf collared
Elongation: Stem elongation		
E0	2.0	Onset of stem elongation
E1	(1/N) + 1.9	First node palpable/visible
E2	(2/N) + 1.9	Second node palpable/visible
En	(n/N) + 1.9	Nth node palpable/visible
Reproductive: Floral development		
R0	3.0	Boot stage
R1	3.1	Inflorescence emergence/first spikelet visible
R2	3.3	Spikelets fully emerged/peduncle not
R3	3.5	Inflorescence emerged/peduncle elongated
R4	3.7	Anther emergence/anthesis
R5	3.9	Post-anthesis/fertilization
Seed development and ripening		
S 0	4.0	Caryopsis visible
S1	4.1	Milk
S2	4.3	Soft dough
S3	4.5	Hard dough
S4	4.7	Endosperm hard/physiological maturity
S5	4.9	Endosperm dry/seed ripe

† Where n equals the event number (number of leaves or nodes) and N equals the number of events within the primary stage (total number of leaves or nodes developed). General formula is $P + (n/N) - 0.1$; where P equals primary stage number (1 or 2 for vegetative and elongation, respectively) and n equals the event number. When $N > 9$, the formula $P + 0.9(n/N)$ should be used.

opment. This system, however, applies only to leaf development through stem elongation and therefore is limited to vegetative development. Simon and Park (1983) modified an earlier system developed by Zadoks et al. (1974) that included eight primary growth stages subdivided into secondary stages. They noted the variability of growth stages in cross-pollinated forage grasses was much larger than in cultivars of self-pollinated cereals. Sanderson (1992) developed a scale for describing development of switchgrass and kleingrass (*Panicum coloratum* L.), but the system is applicable to other grasses. This system was based on the schemes of Haun (1973) and Simon and Park (1983) and was modified to describe 35 stages based on development of leaf, stem, and reproductive structures. The Simon and Park (1983) and Sanderson (1992) systems are complex, detailed, and precise, which makes them better suited to research than for management purposes.

No single system for quantifying morphological development of herbage grasses has been widely accepted (Sanderson, 1992). The system developed by Moore et al. (1991), however, is applicable to most annual and perennial grasses, and is easily applied in the field. This system contains four primary growth stages for quantifying developmental morphology of established perennial grasses: vegetative, elongation, reproductive, and seed ripening (Moore et al., 1991; Table 4–1).

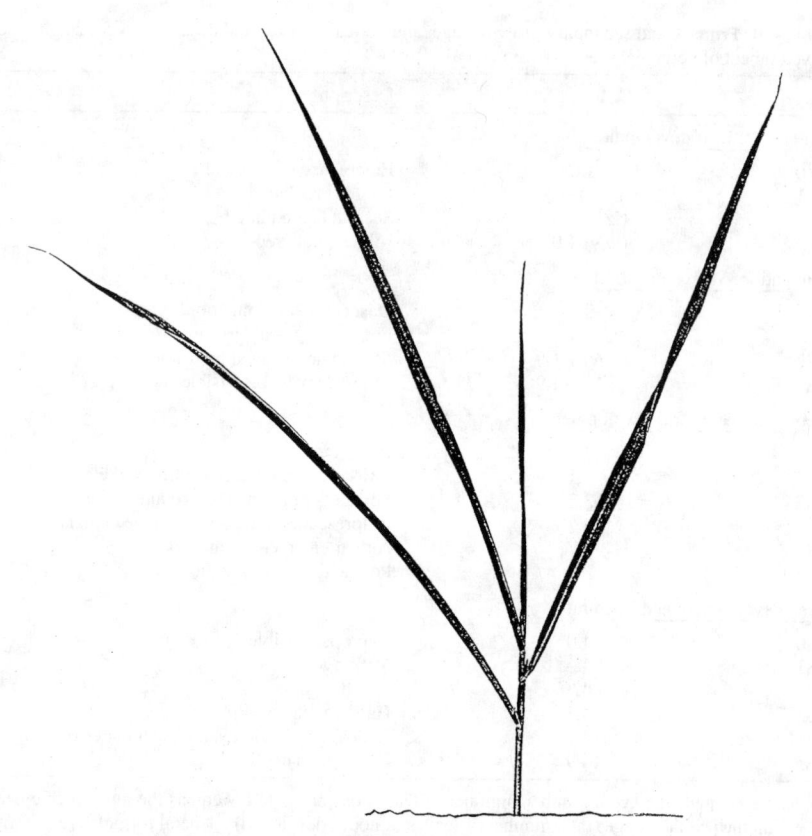

Fig. 4–3. A grass plant that has developed three collared leaves, with a fourth leaf about one-half as long as the third leaf. The numerical scores for five staging systems would be Haun 3.5, Moore et al. V3, Sanderson 3.5, Simon and Park 23, and Zadoks et al. 13 (after Frank, 1996).

Secondary stages within each primary stage describe specific events and are given numerical indices to quantify tiller population development. A representative sample of tillers is collected from the sward to determine the mean growth index for the tiller population based on MSC or MSW (Kalu & Fick, 1981). A grass tiller with three collared leaves and the fourth leaf extended about one-half as long as the third leaf is presented to compare the terminology of five of the staging systems (Fig. 4–3).

Developmental Morphology of Warm-Season Grasses

The Sanderson (1992) system has been used to quantify developmental morphology of 'Alamo' switchgrass and 'Selection 75' kleingrass. Kleingrass flowered earlier and its rate of morphological development was at least twice as fast as switchgrass (Sanderson, 1992). Kleingrass development was nearly 40% slower in summer than in spring, whereas switchgrass developed at the same rate during both periods. Kleingrass developed morphologically at a linear rate during both growth periods. Switchgrass developed morphologically at a linear rate during both periods in the first year of the study, but developed in a sigmoidal pattern in the second year with three distinct phases, which were closely related to temperature and rainfall patterns. Switchgrass developed most rapidly during relatively cool and wet periods. This system described kleingrass and switchgrass development well, but reproductive stages were sometimes subjective.

The Moore et al. (1991) classification system was used to quantify the relationship between developmental morphology of prairie sandreed [*Calamovilfa longifolia* (Hook.) Scribn.] and sand bluestem [*Andropogon gerardii* var. paucipilus (Nash) Fern.] tiller populations with day of the year and GDD (Hendrickson, 1992). Developmental morphology of prairie sandreed and sand bluestem was as closely associated to day of the year as to GDD, with some differences occurring across years. Temporal and spatial variations in morphological development of a native grass sward, however, may make developmental morphology difficult to predict from one growing season to the next (Hendrickson, 1992; Moore & Moser, 1995).

Sanderson et al. (1997) compared the Sanderson (1992) system, the BBCH system (Stauss, 1994), and the Moore et al. (1991) system for quantifying switchgrass morphological development. The primary difference among the staging systems was that the Sanderson and Moore systems were developed for perennial grasses, whereas the BBCH system is a generic scale for describing cereals, oilseed crops, fruits, and vegetable crops. The systems share several common stage descriptors, but have different decimal codes. All staging systems showed the same trend in morphological development with time (Fig. 4–4). They concluded, however, the BBCH system would require substantial modification to adapt to perennial grasses. Additionally, both the Sanderson and BBCH systems have too many detailed stages for producer use. The Moore system has easily remembered stage codes, is easily applied in the field, and apparently was the staging system best suited for quantifying the developmental morphology of perennial grasses.

The Moore et al. (1991) system was used to quantify developmental morphology of 'Trailblazer' switchgrass and 'Pawnee' big bluestem (Mitchell et al., 1997a). The MSC was linearly related to MSW, but MSC slightly underestimated MSW in both species as indicated by the slope coefficients being <1 (Mitchell et al., 1997a; Fig. 4–5). This indicated MSW increased at a faster rate than MSC, which is consistent with other research with alfalfa (Kalu & Fick, 1981), prairie sandreed and sand bluestem (Hendrickson, 1992), and switchgrass (Redfearn et al., 1997). The MSC is a weighted average of all tillers present, so the contribution of juvenile and mature tillers is equally weighted (Kalu & Fick, 1981); however, the MSW is based on the weight of individual tillers. Since mature switchgrass and big bluestem tillers weigh much more than juvenile tillers, mature tillers affect MSW proportionately more. Since MSC and MSW are highly correlated (Fig. 4–5) and MSC is more rapidly and easily determined, quantified developmental morphology can be presented as MSC.

Switchgrass and big bluestem developmental morphology increased linearly with day of the year across six environments in Nebraska and Kansas (Mitchell et al., 1997a). An example of the MSC for switchgrass and big bluestem grown near Mead, NE, in 1990 are plotted with respect to date in Fig. 4–6. Switchgrass MSC

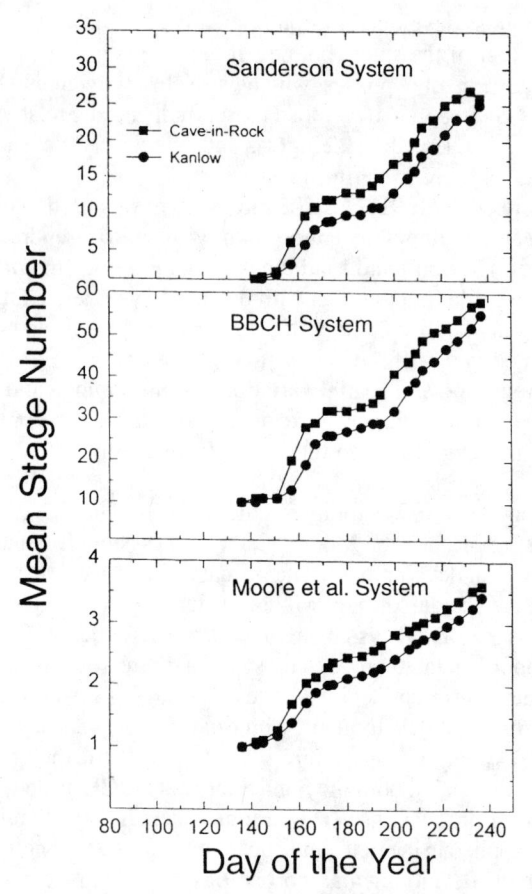

Fig. 4–4. Morphological development of 'Cave-in-Rock' and 'Kanlow' switchgrass grown near Ames, IA, in 1995 and quantified according to the Sanderson (1992), BBCH (Stauss, 1994), and Moore et al., (1991) systems (Sanderson et al., 1997).

advanced linearly with accumulated GDD, whereas big bluestem MSC increased nonlinearly with accumulated GDD in all six environments. Switchgrass MSC was always higher than big bluestem MSC on common days of the year, indicating that switchgrass matured earlier than big bluestem (Fig. 4–6). Switchgrass tillers reached the seed production stage earlier and had a larger proportion of tillers reach the seed production stage than big bluestem, causing the higher MSC for switchgrass. Few big bluestem tillers developed to the seed production stage prior to final harvest.

Variation about the MSC within tiller populations can be estimated by the standard deviation of the MSC (S_{MSC}) and can be used to interpret variability in maturity that exists within tiller populations (Moore et al., 1991). Switchgrass and big bluestem S_{MSC} increased in a similar manner as the growing season progressed at Mead, NE, and Manhattan, KS (Mitchell et al., 1997a; Fig. 4–6). A relatively small S_{MSC} indicates the tiller population is comprised of tillers in similar developmental phases, such as early in the growing season when the tiller population is com-

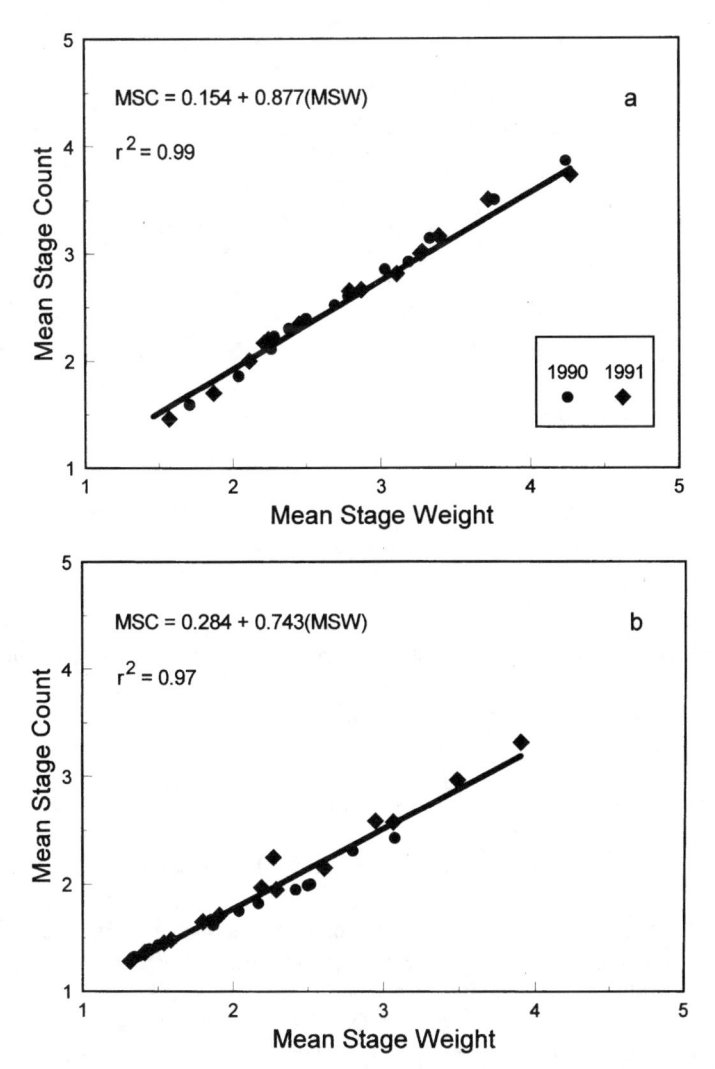

Fig. 4–5. Mean stage count (MSC) and mean stage weight (MSW) relationship for (a) 'Trailblazer' switchgrass and (b) 'Pawnee' big bluestem grown near Mead, NE, in 1990 and 1991 (Mitchell et al., 1997a).

prised solely of vegetative tillers. The increase in S_{MSC} as developmental morphology progresses indicates the tiller population is comprised of an increasingly diverse array of secondary stages, which is illustrated by the tiller demographic data (Fig. 4–7). With intensive grazing management, however, a livestock producer could prevent a large proportion of the tillers from advancing to the reproductive stages, maintaining a small S_{MSC}.

Day of the year is more dependent on photoperiod than GDD. Consequently, the quadratic response of big bluestem to GDD across all environments suggests

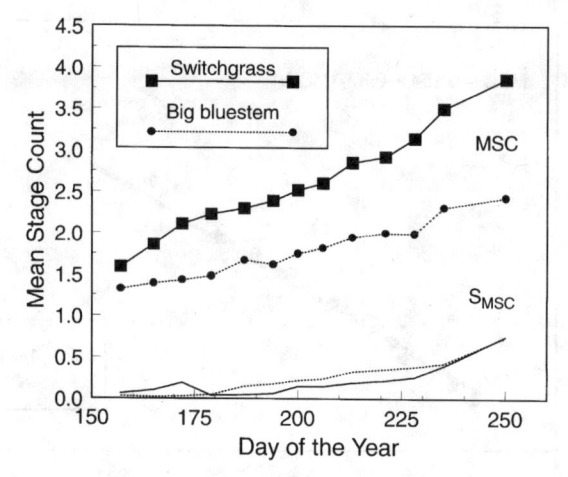

Fig. 4–6. Mean stage count (MSC) and its standard deviation (S_{MSC}) for 'Trailblazer' switchgrass and 'Pawnee' big bluestem grown near Mead, NE, in 1990 (Mitchell et al., 1997a).

morphological development of 'Pawnee' big bluestem is affected more by non-photoperiodic environmental variations than 'Trailblazer' switchgrass. The linear response of switchgrass MSC to day of the year and GDD across all environments indicates morphological development of switchgrass is highly predictable, and suggests that 'Trailblazer' switchgrass is more photoperiod responsive than 'Pawnee' big bluestem. Sensitivity to photoperiod may help explain why more switchgrass tillers advance to the reproductive stages than big bluestem.

Predicting Developmental Morphology

Developmental morphology of perennial warm-season grasses occurs in a predictable manner from one year to the next. A few studies, however, have been conducted to predict developmental morphology of warm-season grasses. In a study conducted in Nebraska and Kansas with 'Trailblazer' switchgrass and 'Pawnee' big bluestem, developmental morphology was successfully predicted (Mitchell et al., 1997a). Developmental morphology of switchgrass and big bluestem was determined at Mead, NE, in 1990 and 1991; these data were used as calibration data to predict switchgrass and big bluestem developmental morphology at Mead, NE, and Manhattan, KS, in 1992 and 1993. The MSC of switchgrass and big bluestem was predicted as a function of day of the year and accumulated GDD. Actual MSC data were plotted against values predicted from the calibration equation.

Swtichgrass developmental morphology was predicted best by a linear day of the year regression equation, which accounted for 96% of the variation in MSC across the four environments (Mitchell et al., 1997a; Fig. 4–8). The high correlation between actual and predicted MSC using the linear day of the year model reiterated the photoperiod sensitivity of switchgrass and indicated that general switchgrass management recommendations for adapted cultivars may be made based on day of the year in the central Great Plains. The Kansas site was approximately 225

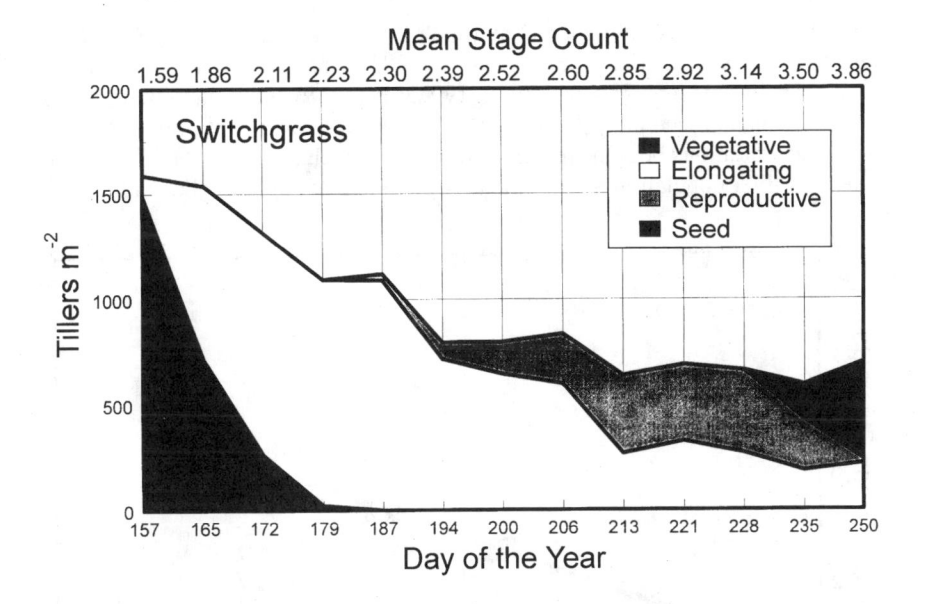

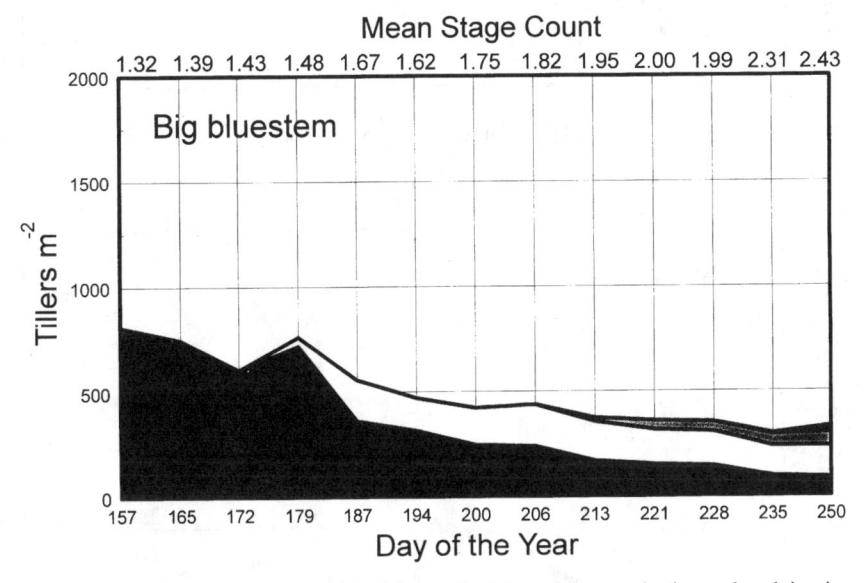

Fig. 4–7. Number of tillers per square meter in vegetative, elongating, reproductive, and seed ripening growth stages for 'Trailblazer' switchgrass and 'Pawnee' big bluestem grown near Mead, NE, during the 1990 growing seasons. The mean stage count is included at the top of the figure as an additional reference for describing the developmental morphology of the tiller population (Mitchell et al., 1997a).

km south of the Nebraska site where the calibration data were collected, which broadens the area of inference of this study. Switchgrass growing in native rangeland, however, may have less uniform and predictable growth patterns, so grazing management based on day of the year should be used cautiously.

Big bluestem developmental morphology was predicted best by a quadratic GDD regression equation, which accounted for 83% of the variation in MSC across the four environments (Mitchell et al., 1997a; Fig. 4–8). The quadratic GDD regression equation accurately predicted MSC across all environments; however, there was more variation about the regression than for switchgrass. The MSC of the big bluestem tiller population in the validation study did not progress beyond the mid

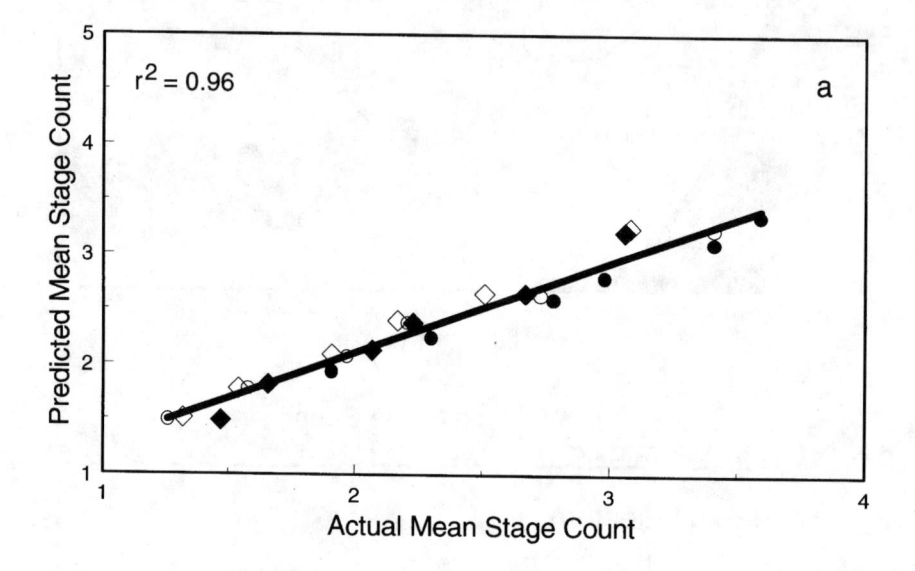

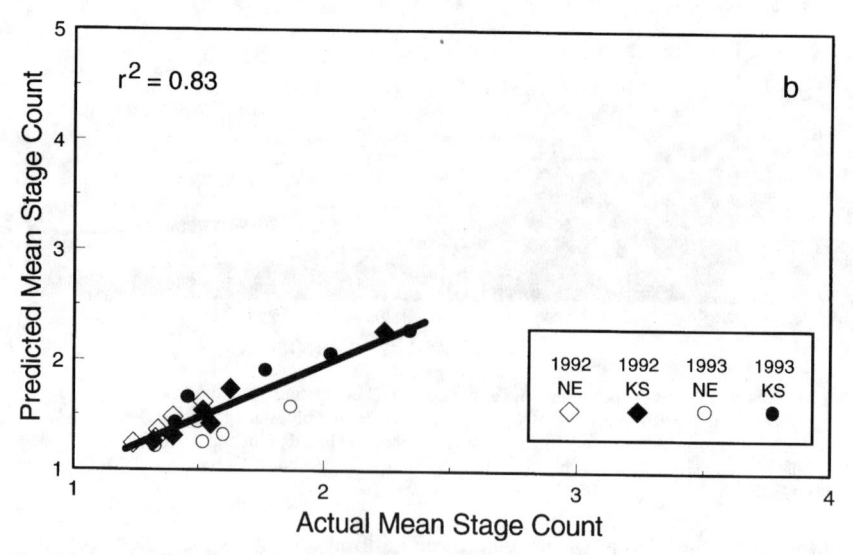

Fig. 4–8. Relationship between mean stage count in four environments and those predicted from (a) a linear day of the year model developed from 'Trailblazer' switchgrass, and (b) a quadratic growing degree day model developed from 'Pawnee" big bluestem grown near Mead, NE, in 1990 and 1991 (Mitchell et al., 1997a).

elongation stages of development, minimizing the range of maturity for prediction. Additionally, few individual tillers developed inflorescences. The good correlation between actual and predicted MSC using the quadratic GDD model indicates the feasibility of making general big bluestem management recommendations across a wide geographic region based on accumulated GDD. Morphological development of switchgrass and big bluestem contrasts, with switchgrass development expressing a more determinate growth habit controlled by photoperiod, whereas big bluestem development is controlled by GDD.

TILLER DYNAMICS

Apical dominance traditionally has been used to explain tiller initiation in perennial grasses, assuming removal of apical meristems by defoliation released axillary buds from hormonal inhibition by auxin, which stimulated tiller initiation (Murphy & Briske, 1992). Although the concept of tiller inhibition by auxin has been used to explain tillering responses, it is no longer considered a valid independent mechanism of physiological control (Nelson, 1996). Cytokinin, a growth regulator produced primarily in the roots, is considered to be associated with tiller stimulation. Auxin production in the shoot apex likely inhibits the synthesis or use of cytokinin in axillary buds that inhibits axillary bud growth (Murphy & Briske, 1992). The auxin/cytokinin ratio in the grass plant may be influential in the developmental morphology of warm-season grasses. This suggests that root growth or root activity may play an important role in the grass tillering process (Nelson, 1996).

Grasses are efficient forage producers because of the location of the meristematic tissue, growth habits of the plant, and the ability of the plant to tiller (Rechenthin, 1956). The number of live tillers within a plant or per unit area is determined by the seasonality of tiller recruitment in relation to tiller longevity (Briske, 1991). The longevity of individual tillers, however, does not exceed 2 yr in many grass species (Briske, 1991). Perennation of established grass swards occurs through asexual reproduction of new tillers or rhizomes from axillary buds (Waller et al., 1985). Tiller density is controlled by the rate of recruitment of new tillers and the mortality of existing tillers (Langer et al., 1964; Briske, 1991). Often, there is high tiller natality and high tiller mortality coinciding with flowering in perennial grasses (Matthew et al., 1993).

The grass sward is a population of tillers that responds to environmental stimuli. The structure of the tiller population changes with time, reflecting the seasonal variability in developmental morphology of individual tillers (Moore & Moser, 1995). Principles of plant demography can be applied with respect to time to describe the variation in developmental morphology within tiller populations. Additionally, canopy measurements such as LAI can be evaluated to determine the changes in canopy architecture during the growing season.

Tiller Demographics

Tiller density represents the pool of meristematic tissues from which growth may occur and therefore, provides the basis for potential productivity (Murphy &

Briske, 1992). Tiller demographic analyses of switchgrass and big bluestem tiller populations are presented in Fig. 4–7 in which the number of tillers per square meter in the vegetative, elongating, reproductive (seedhead development), and seed ripening primary growth stages are plotted with respect to time (Mitchell et al., 1997a). Maximum tiller density for switchgrass and big bluestem occurred on Day 157 and decreased by an average of 9.4 and 5.1 tillers m^{-2} d^{-1} for switchgrass and big bluestem, respectively. Tiller density decreased as morphological development increased for both species. Switchgrass tiller density was nearly twice as high as big bluestem tiller density on common days of the year throughout the entire growing season. Cuomo (1992), however, reported tiller density in July for big bluestem was 95 and 33% higher than switchgrass and indiangrass, respectively.

Approximately 94% of the switchgrass tillers were vegetative at the first sampling date, and MSC was about 1.6 (Fig. 4–7); however, by Day 187, 4-wk after initial sampling, MSC increased to 2.3 with no vegetative tillers present and at least 97% of the tillers were elongating. No vegetative tillers were present throughout the remainder of the sampling period. Nearly 70% of the surviving switchgrass tillers advanced to the reproductive and seed ripening stages.

The large proportion of elongating, reproductive, and seed ripening tillers after Day 172 (21 June) reinforces the importance of grazing switchgrass in late spring and early summer (Fig. 4–7). Improper timing of switchgrass grazing results in low animal performance and may be detrimental to stand persistence (Anderson and Matches, 1983). Grazing of switchgrass should begin when a large proportion of the tillers are vegetative to maximize tiller regrowth, tiller recruitment, and livestock use. If initial switchgrass grazing is delayed until Day 172, livestock still would have the opportunity to selectively graze a limited number of vegetative tillers; however, regrowth from most elongated tillers would be limited following defoliation. If initial grazing is delayed until after Day 187, 100% of the tillers presented to the grazing animal will have elongated, exposing the apical meristem to removal by grazing, potentially decreasing regrowth potential and plant vigor the following spring (Waller et al., 1985).Grazing when a large proportion of tillers are in the elongating and reproductive stages, however, may be used to open the canopy and encourage recruitment of new tillers if soil water is adequate.

At the first four sampling dates, all big bluestem tillers were vegetative, and the MSC ranged from 1.3 to 1.5 (Fig. 4–7). In a 9-wk period from Day 157 to Day 221, big bluestem MSC advanced to only 2.0 with 42% of the tillers still in the vegetative stage. On Day 250 (7 September) visual observation would have described the population as fully headed; however, only 29% of the big bluestem tillers reached the reproductive and seed ripening stages on Day 250. The low MSC for big bluestem throughout the growing season reflected the low proportion of tillers reaching reproductive and seed ripening stages in the population.

Big bluestem matures later in the growing season than switchgrass and is generally recommended as a more productive forage source for mid-summer grazing than switchgrass in the central Great Plains (Moser & Vogel, 1995). The large percentage of vegetative tillers prior to Day 206 (25 July) illustrated why big bluestem is better suited to mid-summer grazing than switchgrass. The large proportion of vegetative tillers throughout the growing season indicated that grazing livestock would have the opportunity to select for higher quality forage from less mature tillers. Additionally, tiller regrowth following grazing would be maximized. Most of the defoliated tillers would be in the vegetative or early elongating stages and a majority of the apical meristems of these tillers would remain intact following defoliation, maintaining regrowth potential of the defoliated tillers. If close grazing occurred when a majority of tillers had elevated apical meristems, regrowth potential would be slower.

Although switchgrass and big bluestem are photoperiod sensitive, uniform tiller advancement toward maturity within a species was not observed. Management of vegetative tillers is crucial in the first growth of switchgrass and big bluestem. The proportion of vegetative tillers declined rapidly in switchgrass, but big bluestem maintained a sizeable proportion of vegetative tillers throughout the growing season (Fig. 4–7), which may indicate that overgrazing would be more detrimental to switchgrass than to big bluestem.

Leaf Area Index

Direct LAI measurement for perennial herbages is a time-consuming process; however, recent technological advances have made LAI estimation possible with indirect optical measurements that are highly correlated with direct measurements (Welles & Norman, 1991). The LAI for 'Trailblazer' switchgrass and 'Pawnee' big bluestem was measured throughout the growing season in Nebraska (Mitchell et al., 1998). Switchgrass and big bluestem LAI increased at similar rates (Fig. 4–9). A maximum LAI of 4.9 was reached by switchgrass and 5.8 by big bluestem on Day 185. These results were similar to those of Redfearn et al. (1997) who reported a maximum LAI of 5.5 for tiller populations of eight switchgrass genotypes. The higher LAI for big bluestem likely occurred in response to the high proportion of vegetative tillers present throughout the growing season that also caused big bluestem MSC to remain low (Fig. 4–7). Day of the year accounted for at least 95 and 84% of the variability in LAI for switchgrass and big bluestem, respectively.

SUMMARY

In this chapter, we have attempted to present recent information concerning developmental morphology and tiller dynamics of warm-season grass swards. Much of this information focused on switchgrass and big bluestem grown in monocultures. Other species, other cultivars within these species, or mixed species stands may exhibit different environmental responses to achieve reproductive success or competitive advantage. A majority of the specifics in this chapter, however, will apply conceptually to other species and cultivars. Differences in developmental morphology exist between species. Managing switchgrass like big bluestem will not be as successful as managing switchgrass according to its own characteristics. Taking into consideration the differences that exist among species will help improve management. A uniform scale for determining developmental morphology of perennial grasses must be adopted to standardize research results and facilitate com-

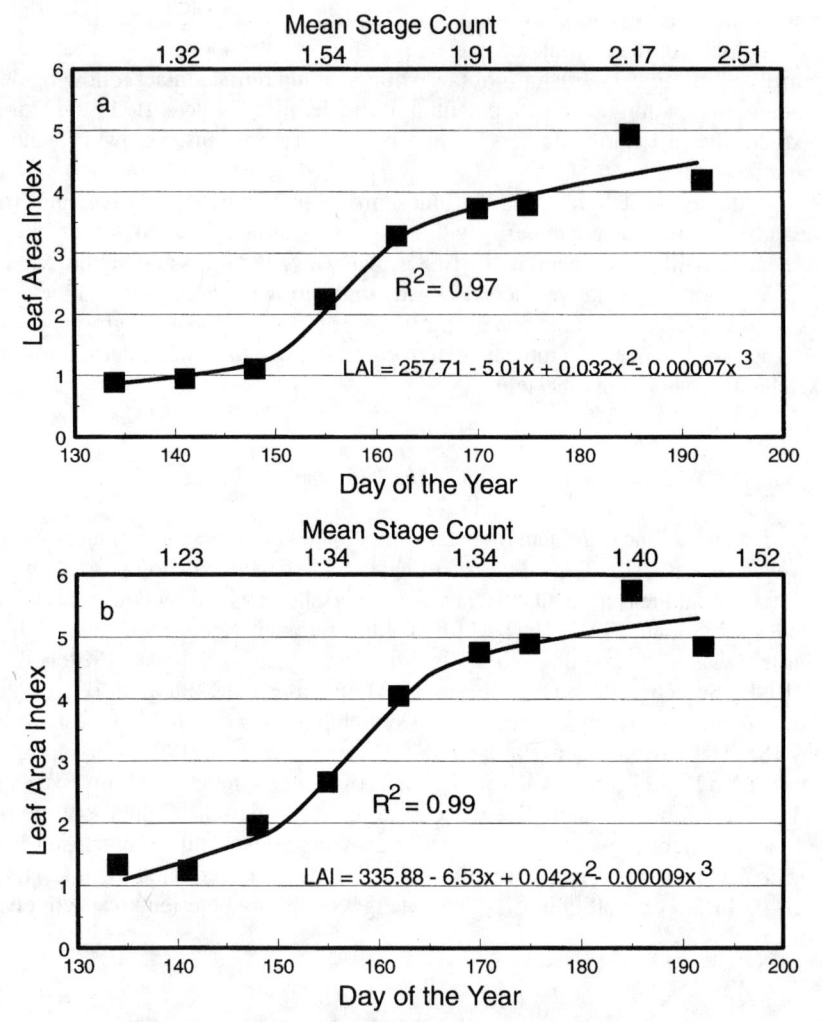

Fig. 4–9. Leaf area index for (a) 'Trailblazer' switchgrass and (b) 'Pawnee' big bluestem grown near Mead, NE, during the 1992 growing season. The mean stage count is included at the top of the figure as an additional reference for describing the developmental morphology of the tiller population (Mitchell et al., 1998).

munication of management recommendations. We suggest adoption of the Moore et al. (1991) system, due to the ease of applying the system in the field and communicating the results to both producers and scientists. Future research should focus on the relationships between developmental morphology and grazing management. Specific research should be conducted to determine the grazing readiness, hay quantity, and hay quality of warm-season grasses based on a uniform scale of developmental morphology. Additionally, determining at what stage of developmental morphology tillering occurs and the impacts of defoliation on developmental morpholgy for important pasture species is crucial.

REFERENCES

Allard, G., C.J. Nelson, and S.G. Pallardy. 1991. Shade effects on growth of tall fescue: I. Leaf anatomy and dry matter partitioning. Crop Sci. 31:163–167.

Allard, H.A., and M.W. Evans. 1941. Growth and flowering of some tame and wild grasses in response to different photoperiods. J. Agric. Res. 62:193–228.

Anderson, B., and A.G. Matches. 1983. Forage yield, quality, and persistence of switchgrass and caucasian bluestem. Agron. J. 75:119–124.

Benedict, H.M. 1941. Effect of day length and temperature on the flowering and growth of four species of grasses. J. Agric. Res. 61:661–672.

Briske, D.D. 1991. Developmental morphology and physiology of grasses. p. 85–108. *In* R.K. Heitschmidt and J.W. Stuth (ed.) Grazing management: an ecological perspective. Timber Press, Portland, OR.

Cuomo, G.J. 1992. Burning and defoliation effects on the vigor and productivity of three warm-season grasses. Ph.D. diss. Univ. of Nebraska, Lincoln.

Dahl, B.E., and D.N. Hyder. 1977. Developmental morphology and management implications. p. 257–290. *In* R.E. Sosebee (ed.) Rangeland plant physiology. Soc. for Range Manage., Denver, CO.

Esau, K. 1960. Anatomy of seed plants. John Wiley, New York.

Fehr, W.R., and C.E. Caviness. 1977. Stages of soybean development. Iowa Coop. Ext. Serv. Spec. Rep. 80. Iowa State Univ., Ames.

Frank, A. 1996. Evaluating grass development for grazing management. Rangelands 18:106–109.

Gillen, R.L., and A.L. Ewing. 1992. Leaf development of native bluestem grasses in relation to degree-day accumulation. J. Range Manage. 45:200–204.

Haun, J.R. 1973. Visual quantification of wheat development. Agron. J. 65:116–119.

Hendrickson, J.R. 1992. Developmental morphology of two Nebraska Sandhills grasses and its relationship to forage quality. M.S. thesis. Univ. of Nebraska, Lincoln.

Hyder, D.N. 1972. Defoliation in relation to vegetative growth. p. 302–317. *In* V.B. Youngner and C.M. McKell (ed.) The biology and utilization of grasses. Academic Press, New York.

Kalu, B.A., and G.W. Fick. 1981. Quantifying morphological development of alfalfa for studies of herbage quality. Crop Sci. 21:267–271.

Langer, R., S. Ryle, and O. Jewiss. 1964. The changing plant and tiller populations of timothy and meadow fescue swards: I. Plant survival and the pattern of tillering. J. Appl. Ecol. 1:197–208.

Matthew, C., C.K. Black, and B.M. Butler. 1993. Tiller dynamics of perennation in three herbage grasses. p. 141–143. *In* M.J. Baker, et al. (ed.) Proc. 17th Int. Grassl. Congr., Palmerston North, NZ. 8–21 February 1993.

Mauseth, J.D. 1988. Plant anatomy. Benjamin/Cummings Publ. Co., Menlo Park, CA.

Mitchell, R.B. 1995. Developmental morphology and forage quality relationships in perennial forage grasses. Ph.D. diss. Univ. of Nebraska, Lincoln.

Mitchell, R.B., K.J. Moore, L.E. Moser, J.O. Fritz, and D.D. Redfearn. 1997a. Predicting developmental morphology in switchgrass and big bluestem. Agron. J. 89:827–832.

Mitchell, R.B., K.J. Moore, L.E. Moser, and D.D. Redfearn. 1998. Tiller demographics and leaf area index of four perennial grasses. Agron. J. 90:47–53.

Mitchell, R.B., L.E. Moser, and K.J. Moore. 1997b. Relationships of visual and quantitative methods of grass sward development. p. 7-7–7-8. *In* Proc. 18th Int. Grassl. Congr., Winnipeg, Manitoba, Canada. 8–19 June 1997.

Moore, K.J., and L.E. Moser. 1995. Quantifying developmental morphology of perennial grasses. Crop Sci. 35:37–43.

Moore, K.J., L.E. Moser, K.P. Vogel, S.S. Waller, B.E. Johnson, and J.F. Pedersen. 1991. Describing and quantifying growth stages of perennial forage grasses. Agron. J. 83:1073–1077.

Moser, L.E., and K.P. Vogel. 1995. Switchgrass, big bluestem, and indiangrass. p. 409–420. *In* R.F Barnes et al. (ed.) Forages: An introduction to grassland agriculture. 5th ed. Iowa State Univ. Press, Ames.

Murphy, J.S., and D.D. Briske. 1992. Regulation of tillering by apical dominance: Chronology, interpretive value, and current perspectives. J. Range Manage. 45:419–429.

Nelson, C.J. 1996. Physiology and developmental morphology. p. 87–125. *In* L.E. Moser et al. (ed.) Cool-season forage grasses. ASA, CSSA, and SSSA, Madison, WI.

Nelson, C., and L. Moser. 1994. Plant factors affecting forage quality. p. 115–154. *In* G.C. Fahey, Jr. et al. (ed.) Forage quality, evaluation, and utilization. ASA, CSSA, and SSSA, Madison, WI.

Phillips, T.G., J.T. Sullivan, M.E. Loughlin, and V.G. Sprague. 1954. Chemical composition of some forage grasses: I. Changes with plant maturity. Agron. J. 46:361–369.

Rechenthin, C.A. 1956. Elementary morphology of grass growth and how it affects utilization. J. Range Manage. 9:167–170.

Redfearn, D., K. Moore, K. Vogel, S. Waller, and R. Mitchell. 1997. Canopy architecture and morphology of switchgrass populations differing in forage yield. Agron. J. 89:262–269.

Salisbury, F.B., and C.W. Ross. 1985. Plant physiology. Wadsworth Publ. Co., Belmont, CA.

Sanderson, M.A. 1992. Morphological development of switchgrass and kleingrass. Agron. J. 84:415–419.

Sanderson, M.A., C.P. West, K.J. Moore, J. Stroup, and J. Moravec. 1997. Comparison of morphological development indexes for switchgrass and bermudagrass. Crop Sci. 37:871–878.

Simon, U., and B.H. Park. 1983. A descriptive scheme for stages of development in perennial forage grasses. p. 416–418. *In* J.A. Smith and V.W. Hays (ed.) Proc. 14th Int. Grassl. Congr., Lexington, KY. 15–24 June 1981. Westview Press, Boulder, CO.

Stauss, R. 1994. Compendium of growth stage identification keys for mono- and dicotyledonous plants. Extended BBCH scale. Ciba-Geigy AG, Basel, Switzerland.

Vallentine, J.F. 1990. Grazing management. Academic Press, San Diego, CA.

Vanderlip, R.L. 1972. How a sorghum plant develops. Kansas Coop. Ext. Serv. Res. Bull. C-447. Kansas State Univ., Manhattan.

Waller, S.S., L.E. Moser, and P.E. Reece. 1985. Understanding grass growth: The key to profitable livestock production. Trabon Printing Co., Kansas City, MO.

Walton, P.D. 1983. Production and management of cultivated forages. Reston Publ. Co., Reston, VA.

Welles, J.M., and J.M. Norman. 1991. Instrument for indirect measurement of canopy architecture. Agron. J. 83:818–825.

Wilson, J.R. 1976. Variation of leaf characteristics with level of insertion on a grass tiller: I. Development rate, chemical composition and dry matter digestibility. Aust. J. Agric. Res. 27:343–354.

Zadoks, J.T., T.T. Chang, and C.F. Konzak. 1974. A decimal code for the growth stages of cereals. Weed Res. 14:415–421.

5 Growing Legumes in Mixtures with Warm-Season Grasses

J. Ronald George, Kevin M. Blanchet, and Randall M. Gettle

Iowa State University
Ames, Iowa

Native warm-season grasses have the potential to provide a large supply of high quality forage during the hot and often dry midsummer months of June, July, and August. Switchgrass (*Panicum virgatum* L.) and big bluestem (*Andropogon gerardii* Vitman) were two major components of the original tall-grass prairie in North America (Weaver, 1968). Although present throughout most of the USA except the far western states, these two grasses were most prominent in this tall-grass prairie region, with big bluestem accounting for up to 80% of the vegetation in some sites. Both have been seeded for summer pasture and hay use, but switchgrass seems to have been more popular during recent years because of lower seed cost and smooth seed characteristics that makes seeding and uniform stand establishment much easier to accomplish.

Roundtree et al. (1974) reported that 75 to 80% of switchgrass dry matter yield is produced from June through August in Missouri, compared with only 40 to 45% of yield for tall fescue (*Festuca arundinacea* Schreb.). This is a time when many cool-season grasses are normally less productive. During this period, switchgrass can produce greater yields than most cool-season grasses (Sharp & Gates, 1986), and greater average daily gains in beef steers (*Bos taurus*) (Wedin & Fruehling, 1977, 1980).

NITROGEN REQUIREMENT

It is generally recognized that cool-season forage grasses have a relatively high N requirement for high yields of quality forage. Warm-season grasses also require relatively large amounts of N for high yields of high-quality forage (Hall et al., 1982; George & Hall, 1983; Smith, 1979). In Iowa, Smith (1981) reported forage yield increases of 95, 148, 187, and 216% when switchgrass was fertilized with 75, 150, 225, and 300 kg N ha^{-1}, respectively, compared with 0 N.

Forage legumes could provide N for warm-season grasses if compatible legumes could be identified that establish and persist in mixed swards, yet not be too competitive with the grass component. These legumes could provide symbiotic N for associated warm-season grasses, improve forage quality, and extend the grazing season.

Copyright © 2000. Crop Science Society of America and American Society of Agronomy, 677 S. Segoe Rd., Madison, WI 53711, USA. *Native Warm-Season Grasses: Research Trends and Issues.* CSSA Special Publication no. 30.

LEGUME–GRASS MIXTURES

Grass swards have benefitted from the addition of legumes when grown as mixtures, with productivity equal to N fertilization (Mallarino & Wedin, 1990; Barnett & Posler, 1983; Wedin et al., 1965).

Botanically stable pasture mixtures are difficult to maintain because well-adapted grass and legume forages may not always make satisfactory companion species. Differences in growth habit, seasonal growth, and relative maturity may lead to incompatibilities between many grasses and legumes when grown together. Because of these differences, persistence of either the grass or legume usually declines over time as one component dominates.

Numerous studies have indicated that a greater proportion of legume to associated grass in the mixture is necessary for optimum yields without N fertilization. In mixtures with cool-season grasses, percentage legume present in high-yielding mixtures was approximately 80% for alfalfa (*Medicago sativa* L.) (Jung et al., 1991), 67% for red clover (*Trifolium pratense* L.) (Mallarino & Wedin, 1990), and 70% for birdsfoot trefoil (*Lotus coniculatus* L.) (Jones et al., 1988).

Blaser et al. (1956) also suggested that establishing and maintaining a desired botanical composition is difficult because the perennial grass and legume components react differently to environmental conditions that affect growth and persistence. Jackobs (1963) stated that the proportion of cool-season grass in legume-grass mixtures usually increases each year, and that supplemental N-fertilization may be needed to produce satisfactory yields by the fourth year. Olsen et al. (1981) reported, however, that birdsfoot trefoil, red clover, and alfalfa tended to dominate cool-season legume–grass mixtures in Illinois.

Many studies have been conducted and reported on cool-season legume–grass forage mixtures. In mixed legume–grass swards, legume persistence is affected by differences in legume and grass competiveness. Either component may dominate the mixture. Marten et al. (1963) observed no difference in alfalfa–smooth bromegrass (*Bromus inermis* Leyss.) yield among stands containing from 18 to 108 alfalfa plants m^{-2}. Birdsfoot trefoil grown in association with Kentucky bluegrass (*Poa pratensis* L.) produced greatest yields at 49 to 107 plants m^{-2} (Taylor et al., 1973).

Well-adapted species may not always be satisfactory companion species when grown in mixtures. Alfalfa dominated mixtures with reed canarygrass (*Phalaris arundinacea* L.) in Iowa (Jones et al., 1988). When established with alfalfa in mixtures, the grass declined in 3 yr to nearly no yield contribution. Birdsfoot trefoil, however, was less competitive, but still dominated the grass. In eastern Kansas, Barnett and Posler (1983) observed alfalfa, red clover, birdsfoot trefoil, and crownvetch (*Coronilla varia* L.) to be compatible in mixtures with cool-season grasses. Mays and Evans (1972) reported in Alabama, however, that crownvetch did not persist in mixtures with tall fescue or orchardgrass (*Dactylis glomerata* L.). Differences in growth habit, seasonal growth, management, and relative maturity may cause problems in maintaining a productive legume–grass sward.

Legumes grown in combination with cool-season grasses can provide symbiotic N to associated grass (Brophy et al., 1987; Farnham & George, 1993, 1994a,b,c; Heichel & Henjum, 1991; Mallarino et al., 1990) and improve total yields

(George, 1984; Nichols & Johnson, 1969). Many of the yield and quality benefits of mixtures may be attributed to N transfer from legume to associated grass. Improved seasonal distribution of forage from mixtures of species that differ in their annual growth patterns also may be beneficial, especially for regions that normally have a shortfall in forage supply during summer. Low production of forage during midsummer is a major limitation of many cool-season species, which ultimately limits productivity of livestock systems.

LEGUME RENOVATION OF ESTABLISHED GRASS

One option for forage-livestock producers is to renovate grass pasture with forage legumes. Both interseeding and frost-seeding (overseeding into an unprepared seedbed while the soil is frozen in northern latitudes, or as the grass sod becomes less productive in southern latitudes) exist as alternatives for legume renovation.

With proper soil moisture and plant cover, alfalfa, red clover, and birdsfoot trefoil can be successfully established into a tall fescue sod (Olsen et al., 1981). Interseeding legumes into grass reduces potential risk of stand failure when compared with frost-seeding. No-till vs. broadcast seeding resulted in greater legume establishment into orchardgrass (Byers & Templeton, 1988) and kleingrass (*Panicum coloratum* L.) (Dovel et al., 1990). In western Canada, Waddington (1992) found best legume seedling emergence in native rangeland when using a heavy drill equipped with depth-controlled double-disk openers, a soil-firming device, and coulters that penetrated residue and hard ground.

Although somewhat more risky than interseeding, especially if not done timely, a simple and low-cost establishment method is to broadcast seed on the soil surface in late winter (frost-seeding in northern USA latitudes, or late summer or fall seeding in more southern latitudes). Soil movement by freezing and thawing, and by rainfall, generally provides adequate seed–soil contact. Establishment by frost-seeding is most successful when the soil surface is honeycombed with ice crystals (Decker & Taylor, 1985). Legumes have been successfully frost-seeded into cool-season grasses in the Midwest (George, 1984; Ahlgren et al., 1944; Graber, 1927), and more recently, into warm-season grass (Jung et al., 1985). Recommended steps for frost-seeding are: (i) minimize competition from existing sod or residue, (ii) correct or maintain proper fertility and pH, (iii) provide good seed–soil contact, and (iv) use suitable management during and after establishment (Charles, 1962). Dowling et al. (1971) recommended leaving moderate amounts of residue (<3.3 Mg ha^{-1}) to increase surface-soil moisture and for restraining the seed to encourage radicle entry into the soil. Too much residue, however, may suspend the seed in vegetation, or may provide too much shade to small seedlings.

LEGUME ESTABLISHMENT AND PERSISTENCE

Nelson and Larson (1984) reported that 40 to 50% legume emergence and eventual establishment is excellent, whereas a more realistic establishment of 33%

was reported in Northeast Regional Publication 42 (1963). Blaser et al. (1956) assigned legumes the following relative seedling vigor scores based on emergence and rates of seedling growth: alfalfa (100), sweetclover (*Melilotus* sp.) (86), red clover (63), and birdsfoot trefoil (41).

Legume persistence is often described as the maintenance of plant numbers to provide adequate productivity. Approximately one-half of the original established plants will survive until spring of the second year (NE Reg. Publ. 42, 1963), although greater initial legume densities would probably have greater rates of mortality (Bolger & Meyer, 1983). Blanchet et al. (1995) reported a range of 39 to 101% persistence during the seeding year and 3 to 68% persistence by the second year when 11 legumes were interseeded with a no-till drill into an established stand of switchgrass in Iowa.

LEGUMES WITH WARM-SEASON GRASSES

Previous research has focused primarily on cool-season, legume-grass mixtures. Very limited information exists on the potential role of cool-season forage legumes when grown in mixed swards with perennial, warm-season grasses.

Legumes have been grown successfully with warm-season grasses in the southern USA to extend the growing season (Evers, 1985). Coastal bermudagrass (*Cynodon dactylon* L. Pers.) or Pensacola bahiagrass (*Paspalum notatum* Flugge) was overseeded with arrowleaf (*T. vesiculosum* Savi) or subterranean (*T. subterraneum* L.) clover. The binary mixtures produced yields equal to monoculture grass fertilized with 168 to 252 kg N ha^{-1}. Clovers provided forage 1 to 2 mo earlier, and reduced the May to June forage peak compared with N-fertilized grass. The N-replacement value of the clovers ranged from 127 to 254 kg ha^{-1}. Fertilization of the clover–grass mixtures with 112 kg N ha^{-1} did not improve yield, forage distribution, or N accumulation, but decreased clover presence by 5 to 33%.

Brown and Byrd (1990) reported that forage yield of an alfalfa and bermudagrass mixture was similar to bermudagrass fertilized with 100 to 300 kg N ha^{-1} in Georgia. Alfalfa dominated the mixture, and ranged from a low of 53% of the forage in August of the first year to a high of 100% for spring harvest in subsequent years. In Kentucky, Templeton and Taylor (1975) reported that yields and forage quality of red clover or bigflower vetch (*Vicia grandiflora* W. Koch) with bermudagrass or tall fescue were superior to those of bermudagrass fertilized with 100 kg N ha^{-1}. Bigflower vetch with bermudagrass was somewhat less productive than red clover with bermudagrass. But, bigflower vetch had the advantages of natural reseeding, ease of establishment in grass sod, and provided severe competition to winter annual weeds.

Marten (1989) suggested that warm-season grasses may be incompatible with cool-season legumes because of differences in growth habit, relative maturity, harvest schedules, and poor persistence. Taylor and Jones (1983) observed that alfalfa and red clover were not compatible with switchgrass grown in Kentucky, regardless of cutting management, because these two legumes dominated the mixture after becoming established. Bigflower vetch in this same study, however, persisted well with switchgrass, with the mixture containing 20% legume and having an average annual yield of 9.4 Mg ha^{-1} during a 7-yr period. In Pennsylvania, red clover, birdsfoot trefoil, and bigflower vetch overseeded into established switchgrass yielded >8.0 Mg ha^{-1} annually, with approximately 55% of forage production during July and August when cool-season grasses usually are less productive (Jung et al., 1985). Success varied with grass stand ratings, however, with 30 or 6% red clover for grass stand ratings of 55 or 90%, respectively.

In Missouri, Matches and Mitchell (1976) determined that 'Dawn' birdsfoot trefoil and 'Serala' Sericea lespedeza [*Lespedeza cuneata* (Dum.-Cours.) G. Don] were better adapted for growing with switchgrass or caucasian bluestem [*Bothriochloa caucasia* (Trin.) C.E. Hubb.] than were 'Cody' alfalfa or 'Summit' korean lespedeza [*Kummerowia stipulacea* (Maxim.) Makino] when legumes and grasses were seeded together. Matches and Mowery (1979) successfully established birdsfoot trefoil and red clover into switchgrass and caucasian bluestem using sod seeding, but had erratic results with alfalfa and ladino clover (*T. repens* L.). Legume canopies had to be removed frequently to prevent shading injury and loss of grass stand.

Posler et al. (1993) observed greater yields in Kansas with four of six native legume–switchgrass mixtures compared with nonfertilized switchgrass. Warm-season legumes ranged from 17 to 74% of the botanical composition for these legume–grass mixtures. Stands of the three warm-season grasses, which consisted of switchgrass, sideoats grama (*Bouteloua curtipendula* Michx.), or indiangrass (*Sorghastrum nutans* Nash.), were dominated by warm-season legumes having a prostrate growth habit, but were relatively unaffected by taller legumes.

COMPETITION, GROWTH, AND DEFOLIATION MANAGEMENT

A major concern of some forage scientists is that mixtures of cool-season legumes with warm-season grasses would not be compatible (Marten, 1989). Earlier growth of the legume could deplete soil moisture needed by the grass (probably less critical in higher rainfall regions), and early legume growth could be too competitive with the later growth initiation of warm-season grasses. Timely defoliation of early legume growth by cutting or grazing management may minimize these potential problems.

Bud activity and morphological development are important management considerations for warm-season grasses (Heidemann & Van Riper, 1967; Sims et al., 1971). George and Reigh (1987) reported that the apical meristem of switchgrass typically is below the soil surface up to about 1 June at central Iowa latitude, so early partial defoliation should not seriously affect primary growth and subsequent productivity of the warm-season grass, and would greatly reduce competition to associated legumes during establishment in the mixture when seeded into established grass. It seems logical that early partial defoliation of big bluestem could be delayed until about mid-June, because developmental growth of bluestem typically lags about 2 wk behind that of switchgrass. Additionally, a 5- or 6-wk period of recovery growth for each grass, after early partial defoliation, before summer grazing should allow replenishment of total nonstructural carbohydrates (TNC) reserves to maintain a productive and persistent grass stand (George et al., 1989).

Early, partial defoliation of switchgrass has improved summer herbage quality and total yield of crude protein and digestible dry matter (George & Obermann, 1989; George et al., 1990; Hunczak, 1992). Seasonal dry matter yield in these studies was equal or greater than that for switchgrass that was not partially defoliated in early June. And, use of this rotational grazing, early-defoliation management in a pasture system dramatically improved performance of grazing steers compared with a later initiation of grazing for continuous grazing management (George et al., 1996). Furthermore, this defoliation management would seem to be useful to minimize excessive grass competition during legume seedling growth if the established grass is renovated with forage legumes.

INTERSEEDING AND FROST-SEEDING LEGUMES INTO ESTABLISHED SWITCHGRASS

There are three major potential benefits for legume renovation of established warm-season grasses: (i) replacement of the N fertilizer requirement for monoculture grasses, (ii) increased forage yield, and (iii) improved forage quality. The remainder of this chapter will report the results for cool-season legume renovation of switchgrass by either interseeding (Blanchet et al., 1995; George et al., 1995) or frost-seeding (Gettle et al., 1996a,b) of legumes into established switchgrass. Legume renovation was combined with early June, partial-defoliation management during the establishment year and the first production year.

Legume Establishment

Excellent legume stands were achieved by early June, 10 wk after interseedings were made (Blanchet et al., 1995), with mean densities of 195 and 163 plants m^{-2} for seedings made in 1991 and 1992, respectively. Sweetclover, birdsfoot trefoil, alfalfa, and red clover, when interseeded, generally had the greatest legume plant densities. Crownvetch and hairy vetch (*Vicia villosa* Roth) had the lowest legume plant densities, mostly because of lesser numbers of seeds planted, i.e., percentage of establishment for these two legumes was not greatly different from that for the other legumes. Initial legume plant densities were generally similar or greater than those reported in previous grass renovation studies (Marten et al., 1963; Olson et al., 1981)

Successful legume establishment was achieved by frost-seeding with a mean of approximately 25% of seeds producing established plants by June of the seeding year (Gettle et al., 1996a), compared with about 40 to 50% success for interseeding (Blanchet et al., 1995). Mean success was slightly less than that reported for conventional establishment in Northeast Regional Publ. 42 (1963), but except for the two sweetclovers, was within the 20 to 50% range given by Barnes and Sheaffer (1985) for conventional establishment practices. Plant densities tended to be lower for the sweetclovers compared with birdsfoot trefoil, red clover, alfalfa, and the birdsfoot trefoil-red clover mixture (Fig. 5–1). Mean legume plant densities of 160 and 170 plants m^{-2} were achieved by June of the establishment year for 1991 and 1992 frost-seedings, respectively. Because these densities are similar to that re-

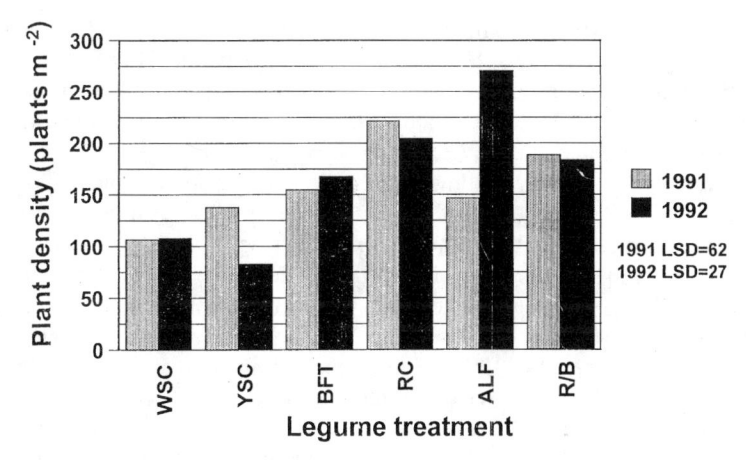

Fig. 5–1. Average plant density in June of the seeding year for legumes frost-seeded into established switchgrass in 1991 and 1992. WSC, white sweetclover; YSC, yellow sweetclover; BFT, birdsfoot trefoil; RC, red clover; ALF, alfalfa; R/B = 50:50 mixture of medium red clover–birdsfoot trefoil.

ported for interseeding (Blanchet et al., 1995), it seems that most cool-season legumes can be successfully established into stands of switchgrass with either seeding method.

Legume Persistence

Most legumes maintained excellent stand densities into October of the establishment year for interseeding (Fig. 5–2), or for frost-seeding, and by June of the second year (Fig. 5–3), although differences among legume species did occur (Blanchet et al., 1995). Mean legume persistence to October of the establishment

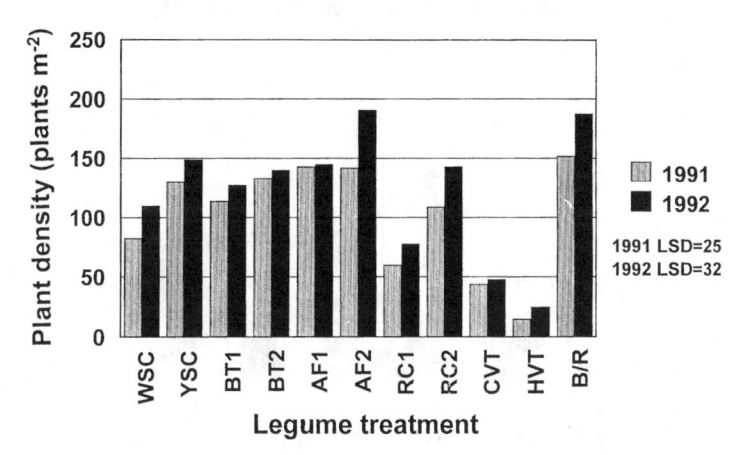

Fig. 5–2. Average plant density in October of the establishment year for legumes interseeded into established switchgrass in 1991 and 1992. Wsc, Polara sweetclover; Ysc, Madrid sweetclover; Bt1, Norcen birdsfoot trefoil; Bt2, Fergus birdsfoot trefoil; Af1, Apollo Supreme alfalfa; Af2, Alfagraze alfalfa; Rc1, Mammoth red clover; Rc2, Redland II red clover; Cvt, Emerald crownvetch; Hvt, Common hairy vetch; B/R, 50:50 mixture of medium red clover–birdsfoot trefoil.

year for interseeding was 52% for 1991 and 78% for 1992 seedings. The somewhat lower initial plant densities for 1992 seedings may have resulted in less within-row competition among legumes during the 1992 establishment year, and greater persistence into October. Greater mortality has been reported for high-plant-density alfalfa (Rowe, 1988).

Severe winter weather conditions and extremely wet spring conditions caused major legume mortality by June of the second year in 1992 seedings for both interseeding (Blanchet et al., 1995) and frost-seeding (Fig. 5–3). Legume species differed in response to these unfavorable weather conditions, with birdsfoot trefoil and Mammoth red clover being most tolerant. Mean legume persistence with interseeding averaged 43% for 1991 seedings and only 26% for weather-damaged stands of 1992 seedings. For 1991 seedings, the trefoil, alfalfa, and red clover cultivars and the birdsfoot trefoil–red clover mixture had excellent persistence of nearly 100% from fall to June of the second year. Thus, with reasonable weather conditions, it seems that many cool-season legumes should persist into the second year when interseeded into mixed swards with established switchgrass.

Mean legume persistence for frost-seeded legumes from June to September of the establishment year was 77% for 1991 seedings and 88% for 1992 seedings (Gettle et al., 1996a). Greater persistence resulted for broadcast frost-seedings, than for row-planted interseedings (Blanchet et al., 1995; Rowe, 1988).

Although differences occurred among legume species for frost-seeding, most had 60 to 85% or greater persistence from June to September, which suggests that switchgrass did not cause serious competition for the establishing cool-season legume plants. By June of the second year, persistence had decreased to 59 and 33% for 1991 and 1992 seedings, respectively, lower for 1992 because of unfavorable weather for good survival of several legume species. Mean legume plant densities of 95 and 55 plants m^{-2} existed by June of the second year. With reasonable weather conditions, legume plant densities for birdsfoot trefoil, red clover, alfalfa, and the birdsfoot trefoil–red clover mixture generally would be more than adequate

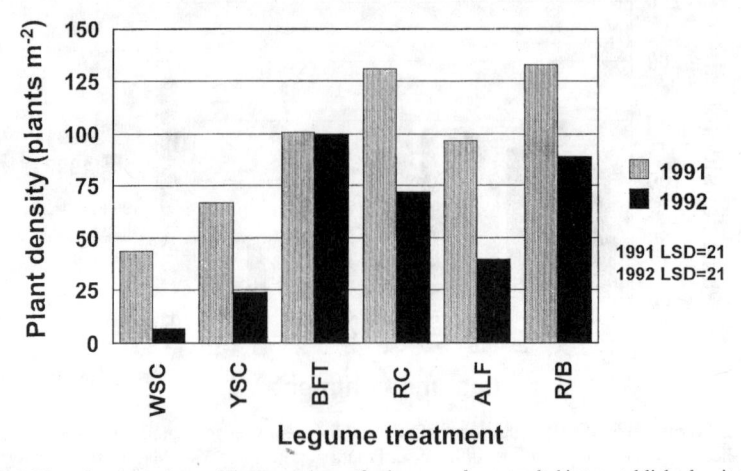

Fig. 5–3. Plant density in June of the second year for legumes frost-seeded into established switchgrass in 1991 and 1992. WSC, white sweetclover; YSC, yellow sweetclover; BFT, birdsfoot trefoil; RC, red clover; ALF, alfalfa; R/B = 50:50 mixture of medium red clover–birdsfoot trefoil.

for a legume–grass mixed sward with either frost-seeding (Gettle et al., 1996a) or interseeding (Fig. 5–4). Sweetclover plant densities were lower than for the other legume species, possibly because of greater insect pressure, especially the sweetclover weevil (*Sitona cylindricollis* Fahraeus). Finally, with the exception of the biennial sweetclovers that had disappeared from the swards by midsummer of the second year, mean persistence of the remaining three frost-seeded legume species and the mixture was 59 and 32% by September, which is essentially identical to mean densities that were observed in June of the second year. Thus, excellent establishment and persistence of most cool-season legumes was achieved for a 2-yr period when frost-seeded or interseeded into established switchgrass.

The generally successful establishment and presistence for legumes that were either interseeded or frost-seeded into established switchgrass probably was influenced greatly by defoliation management that was used during the establishment year and second year in these studies. A flail harvester was used to uniformly defoliate forage to a 15-cm height in early June of the establishment year, to reduce switchgrass canopy competition during early legume establishment, and to simulate early partial defoliation of switchgrass by livestock grazing (George et al., 1996; George & Obermann, 1989, Hunczak, 1992). Forage–livestock producers could manage similar defoliation with short-duration, intensive stocking rates. Forage was harvested again in early June of the second year at a 10-cm height to remove the early growth of established cool-season legumes and to reduce canopy competition of the later growth initiation by switchgrass, and to again simulate early partial defoliation by grazing livestock. Additionally, harvests were made in a similar manner at a 20-cm cutting ht in July of the establishment year, and in July and August of the second year.

It might be important that the established switchgrass variety used in these studies was 'Cave-in-rock', which has a coarser stem and a lesser tiller density than that for varieties like 'Blackwell' or 'Pathfinder'. Thus, it is possible that legume

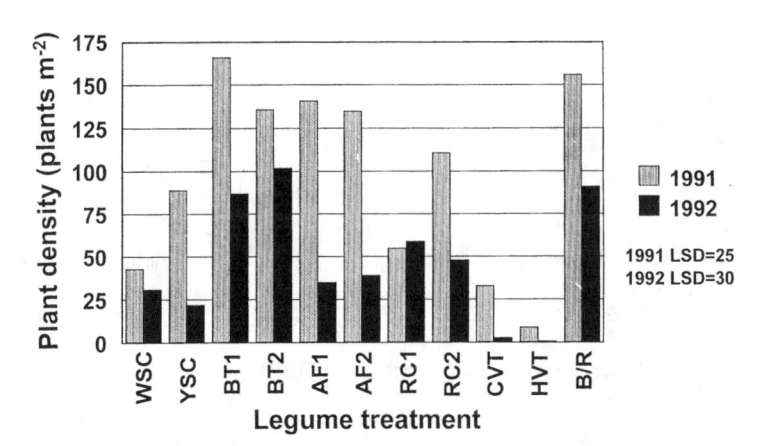

Fig. 5–4. Plant density in June of the second year for legumes interseeded into established switchgrass in 1991 and 1992. Wsc, Polar sweetclover; Ysc, Madrid sweetclover; Bt1, Norcen birdsfoot trefoil; Bt2, Fergus birdsfoot trefoil; Af1, Apollo Supreme alfalfa; Af2, Alfagraze alfalfa; Rc1, Mammoth red clover; Rc2, Redland II red clover; Cvt, Emerald crownvetch; Hvt, Common hairy vetch; B/R, 50:50 mixture of medium red clover–birdsfoot trefoil.

renovation of other switchgrass varieties might give different results compared with those obtained in these studies, or that modification in defoliation management might be necessary to achieve similar success.

Forage Yield

Interseeding or frost-seeding of cool-season legumes into established switchgrass produced similar results in forage yield responses, so only interseeding yields are presented in the figures. Legume renovation did not affect forage yield in June of the establishment year compared with yield for switchgrass that was not fertilized with N, but provided an 8 to 9% greater yield in July (George et al., 1995; Gettle et al., 1996b). Fertilization with N increased forage yield 2.4-fold compared with 0 N or legume renovation (Fig. 5–5). Because legumes require time to establish and contribute to increased forage yield, improve forage quality, and to provide symbiotic N to the associated grass in mixtures, livestock producers should consider renovating only a portion of their warm-season grass pastures each year. Otherwise they should expect to experience a major shortfall in forage supply during legume establishment, in contrast to immediate yield increase of warm-season grass pastures with N fertilization (Hall et al., 1982; George & Hall, 1983; Smith, 1979)

During the second year for 1991 seedings, most legume renovation treatments produced more total forage than grass fertilized with 240 kg N (Fig. 5–6). During the second year for 1992 seedings, most legumes were seriously affected by unfavorable weather, and only birdsfoot trefoil, Mammoth red clover, and the birdsfoot trefoil–red clover mixture had yields that equaled or exceeded that for 240 kg N (Fig. 5–7). Thus, yields of legume–renovated switchgrass were generally greater than for mid- to high-levels of N fertilization, and offer an alternative other than N fertilization. Weather risks with most legumes and the time lag between renovation

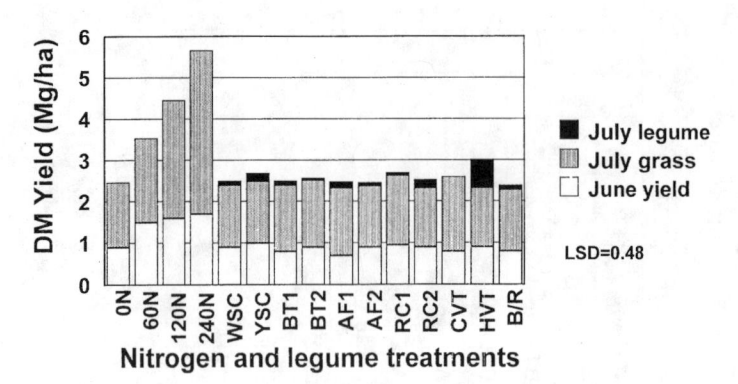

Fig. 5–5. Establishment year upper canopy (>20 cm) total-season dry matter (DM) yields and grass–legume composition averaged for 1991 and 1992 for 11 legume renovation treatments interseeded into established switchgrass compared with N-fertilized and nonfertilized switchgrass. 0N, 60N, 120N, and 240N = kg N ha^{-1}; Wsc, white sweetclover; Yse, yellow sweetclover; Bt1, Norcen trefoil; Bt2, Fergus trefoil; Af1, Apollo Supreme alfalfa; Af2, Alfagraze alfalfa; Rc1, Mammoth red clover; Rc2, 50:50 mixture of medium red clover; Cvt, crownvetch; Hvt, hairy vetch; B/R, 50:50 medium red clover–birdsfoot trefoil.

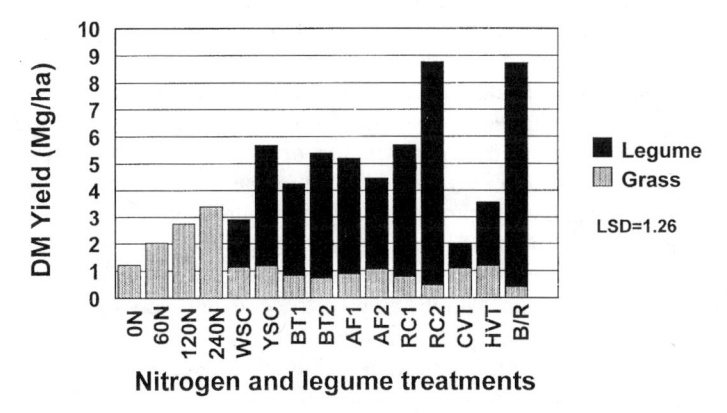

Fig. 5–6. Second-year upper canopy (>20 cm) total-season dry matter (DM) yields harvested in 1992 and grass–legume composition for 11 legume-renovation treatments interseeded into established switchgrass in 1991 and compared with N-fertilized and nonfertilized switchgrass. 0N, 60N, 120N, and 240N = kg N ha^{-1}; Wsc, white sweetclover; Yse, yellow sweetclover; Bt1, Norcen trefoil; Bt2, Fergus trefoil; Af1, Apollo Supreme alfalfa; Af2, Alfagraze alfalfa; Rc1, Mammoth red clover; Rc2, Medium red clover; Cvt, crownvetch; Hvt, hairy vetch; B/R, 50:50 mixture of medium red clover–birdsfoot trefoil.

and increased pasture productivity should be carefully considered by the forage-livestock producer.

Botanical Composition of the Mixed Sward

Adapted interseeded legumes dominated (Fig. 5–6) the herbage of the mixed sward in the second year (George et al., 1995; Gettle et al., 1996b). Mean legume composition for June, July, and August for 1991 interseedings was 84, 70, and 51%

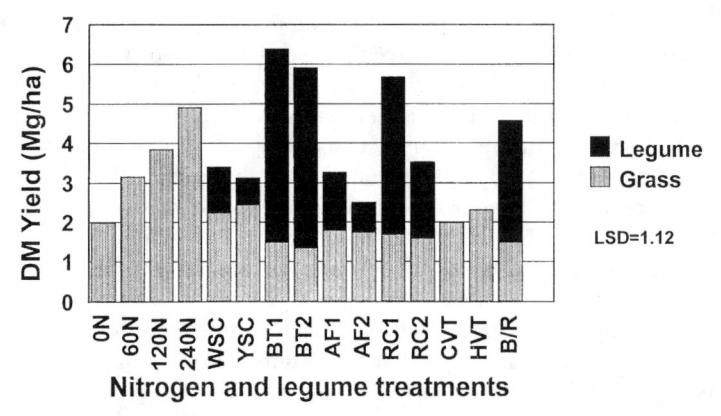

Fig 5–7. Second-year upper canopy (>20 cm) total-season dry matter (DM) yields harvested in 1993 and grass–legume composition for 11 legume treatments interseeded into established switchgrass in 1992 and compared with N-fertilized and nonfertilized switchgrass. 0N, 60N, 120N, and 240 N = kg N ha^{-1}; Wsc, white sweetclover; Yse, yellow sweetclover; Bt1, Norcen trefoil; Bt2, Fergus trefoil; Af1, Apollo Supreme alfalfa; Af2, Alfagraze alfalfa; Rc1, Mammoth red clover; Rc2, Medium red clover; Cvt, crownvetch; Hvt, hairy vetch; B/R, 50:50 mixture of medium red clover–birdsfoot trefoil.

of the total biomass, respectively, and 53, 28, and 27% in weather-damaged stands of 1992 seedings. Legume composition for July and August of the second year declined from June levels, leading to a more even balance of legume–grass forage. Legume renovation improved seasonal distribution of forage by increasing early-season June forage supply compared with that of pure grass, even when fertilized with N. Legume renovation also yielded more forage in spring and summer, and for the season, compared with moderate to heavy N fertilization during the second year, if severe weather did not damage legumes.

Possible alternatives exist other than intensive defoliation management if there is concern that established legumes may be too competitive with warm-season grasses which initiate slower and later spring growth. These include either a reduction in legume seeding rates, or an increase in row width for interseeding of legumes, especially where early defoliation management of the vigorous cool-season legume growth is questionable.

Forage Quality

The two major questions that exist for potential improvement of forage quality by renovation of grass with forage legumes are: (i) did legume renovation improve quality of the total forage, and (ii) did legume renovation improve quality of the grass component in the mixed legume-grass forage?

Establishment Year

Legume renovation by either interseeding (Blanchet, 1994) or frost-seeding (Gettle, 1994) did not improve early June forage protein or IVDMD of either the total mixed sward or the grass component in the seeding year above that for nonfertilized switchgrass. Legume renovation would not be expected to provide additional forage protein or improved forage digestibility only 2 to 3 mo after seedings are made, because legume plants are small and comprised little to none of the forage harvested at a 15-cm cutting ht, and no significant N transfer to the associated grass would be expected at this relatively early stage of legume seedling establishment. In contrast, N fertilization increased forage protein by 22 to 74%, ranging from 111 to 166 g kg^{-1} dry matter, compared with only 86 to 101 g kg^{-1} for interseeded or frost-seeded legume renovation or the N check.

By July of the seeding year, total forage protein concentration above the 20-cm cutting height was equal or greater in legume interseeded plots than that for 120 to 240 kg ha^{-1} N fertilization, and was especially greater for interseeded yellow sweetclover, medium red clover, and hairy vetch swards. Improvement in forage IVDMD also resulted from hairy vetch renovation. Those legumes that establish more quickly and make more first-season growth improved forage quality most during the establishment year. Heichel et al. (1985) reported that first-year red clover stands yielded significantly more forage and fixed significantly more N than for birdsfoot trefoil stands. For frost-seeding, improved forage protein of legume-renovated switchgrass in July of the establishment year was generally similar to that observed for interseeding, and IVDMD was generally greater than that for nonfertilized or N-fertilized switchgrass.

Second Year

For interseeding, Norcen trefoil, Mammoth red clover, and the trefoil-medium red clover mixture produced forage in the July mixed sward that had greater concentrations of protein and IVDMD compared with that for medium to heavy N fertilization of monoculture switchgrass. In August, mixed swards of two trefoil and two alfalfa cultivars produced forage of greater protein and IVDMD concentrations compared with that for moderate to heavy N fertilization for monoculture switchgrass.

For frost-seeding, improved forage quality of the mixed legume-grass sward compared with N fertilization was similar to that observed for interseeded legume renovation. Apparent N transfer from the associated legumes to the companion grass in mixed swards peaked in July and August when maximum switchgrass growth occurred. Protein concentrations of the switchgrass component from mixed legume–grass swards were similar to switchgrass when fertilized with 120 to 240 kg N ha^{-1}. Also, protein concentrations of forage legumes tended to be greater than that for N-fertilized switchgrass in July and August when switchgrass protein was declining. Legume digestibility was about 10% greater than grass digestibility at each harvest. The less leafy, stem elongation characteristic of switchgrass would be expected to result in lower forage quality later in the season. Selective grazing of grass leaf blades by livestock, however, could minimize these quality differences between legumes and grass.

Total season production of protein and digestible dry matter was greatest with N fertilization during the establishment year. In the second year, however, total protein and digestible dry matter yield for mixed legume-grass forage was generally greater or equal to that for N fertilization at 240 kg ha^{-1}.

Thus, legume renovation of established switchgrass seems to offer the potential to increase forage quality modestly by mid to late summer of the establishment year, depending upon the choice of legume. Legume renovation of established switchgrass would probably increase forage yield and quality of the mixed sward during the second year, and replace the need for N fertilization of monoculture switchgrass, when compared with modest to high levels of N fertilization.

REFERENCES

Ahlgren, H.L., M.L. Wall, J. Muckenhirn, and F.V. Burcalow. 1944. Effectiveness of renovation in increasing yields of permanent pastures in southern Wisconsin. J. Am. Soc. Agron. 36:121–131.

Anonymous. 1963. Northeast Reg. Publ. 42. New Jersey Agric. Exp. Stn. Bull. 804.

Barnes, D.K., and C.C. Sheaffer. 1985. Alfalfa. p. 89–97. In M.E. Heath et al. (ed.). Forages: The science of grassland agriculture. 4th ed. Iowa State Univ. Press, Ames.

Barnett, F.L., and G.L. Posler. 1983. Performance of cool-season perennial grasses in pure stands and in mixtures with legumes. Agron. J. 75:582–586.

Blanchet, K.M. 1994. Renovation of established switchgrass by interseeding forage legumes. M.S. thesis. Iowa State University Library, Ames, IA.

Blanchet, K.M., J.R. George, R.M. Gettle, D.R. Buxton, and K.J. Moore. 1995. Establishment and persistence of legumes interseeded into switchgrass. Agron. J. 87:935–941.

Blaser, R.E., T. Taylor, W. Griffeth, and W. Skrdla. 1956. Seedling competition in establishing forage plants. Agron. J. 48:1–6.

Bolger, T.P., and D.W. Meyer. 1983. Influence of plant density on alfalfa yield and quality. p. 37–41. *In* Proc. Am. Forage Grassl. Conf., Lexington, KY. 15–24 June 1981. Westview Press, Boulder, CO.

Brophy, L.S., G.H. Heichel, and M.P. Russelle. 1987. Nitrogen transfer from legumes to grass in a systematic planting design. Crop Sci. 27:753–758.

Brown, R.H., and G.T. Byrd. 1990. Yield and botanical composition of alfalfa–bermudagrass mixtures. Agron. J. 82:1074–1079.

Byers, R.A., and W.C. Templeton, Jr. 1988. Effects of sowing date, placement of seed, vegetation suppression, slugs and insects upon establishment of no-till alfalfa in orchardgrass sod. Grass For. Sci. 43:279–289.

Charles, A.H. 1962. Pasture establishment by surface-sowing methods. Herb. Abstr. 32:175–181.

Decker, A.M., and T.H. Taylor. 1985. Establishment of new seedlings and renovation of old sods. p. 288–297. *In* M.E. Heath et al. (ed.). Forages: The science of grassland agriculture. 4th ed. Iowa State Univ. Press, Ames.

Dovel, R.L., M.A. Hussey, and E.C. Holt.1990. Establishment and survival of Illinois bundleflower interseeded into an established kleingrass pasture. J. Range Manage. 43:153–156.

Dowling, P.M., R.J. Clements, and J.R. McWilliam. 1971. Establishment and survival of pasture species from seeds sown on the soil surface. Aust. J. Agric. Res. 22:61–74.

Evers, G.W. 1985. Forage and nitrogen contributions of arrowleaf and subterranean clovers overseeded on bermudagrass and bahiagrass. Agron. J. 77:960–963.

Farnham, D.E., and J.R. George. 1993. Dinitrogen fixation and nitrogen transfer among red clover cultivars. Can. J. Plant Sci. 73:1047–1054.

Farnham, D.E., and J.R. George. 1994a. Dinitrogen fixation and nitrogen transfer in birdsfoot trefoil–orchardgrass communities Agron. J. 86:690–694.

Farnham, D.E., and J.R. George. 1994b. Harvest management effects on dinitrogen fixation and nitrogen transfer in red clover–orchardgrass mixtures. J. Prod. Agric. 7:360–364.

Farnham, D.E., and J.R. George. 1994c. Harvest management effects on productivity, dinitrogen fixation, and nitrogen transfer in birdsfoot trefoil: orchardgrass communities. Crop Sci. 34:1650–1653.

George, J.R. 1984. Grass sward improvement by frost-seeding with legumes. p. 265–269. *In* Proc. Am. Forage Grassl. Conf., Houston, TX. 23–26 Jan. 1984. Am. Forage and Grassl. Council, Georgetown, TX.

George, J.R., K.M. Blanchet, R.M. Gettle, D.R. Buxton, and K.J. Moore. 1995. Yield and botanical composition of legume-interseeded vs. nitrogen-fertilized switchgrass. Agron. J. 87:1147–1153.

George, J.R., and K.E. Hall. 1983. Herbage quality of three warm-season grasses with nitrogen fertilization. Iowa State J. Res. 58(2):247–259.

George, J.R., R.L. Hintz, K.J. Moore, S.K. Barnhart, and D.R. Buxton. Steer response to rotational or continuous grazing on switchgrass and big bluestem pastures. 1996. p. 150–154. *In* Proc. Am. Forage and Grassl. Conf., Vancouver, BC. 13–15 June 1996. Am. Forage and Grassl. Coun., Georgetown, TX.

George, J.R., and D. Obermann. 1989. Spring defoliation to improve summer supply and quality of switchgrass. Agron. J. 81:47–52.

George, J.R., D.J. Obermann, and D.D. Wolf. 1989. Seasonal trends for non-structural carbohydrates in stem bases of defoliated switchgrass. Crop Sci. 29:1282–1287.

George, J.R., and G.S. Reigh. 1987. Spring growth and tiller characteristics of switchgrass. Can. J. Plant Sci. 67:167–174.

George, J.R., G.S. Reigh, R.E. Mullen, and J.J. Hunczak. 1990. Switchgrass herbage and seed yield and quality with partial spring defoliation. Crop Sci. 30:845–849.

Gettle, R.M. 1994. Renovation of established switchgrass by frost-seeding forage legumes. M.S. thesis. Iowa State Univ., Ames.

Gettle, R.M., J.R. George, K.M. Blanchet, D.R. Buxton, and K.J. Moore. 1996a. Frost-seeding legumes into established switchgrass: establishment, density, persistence, and sward composition. Agron. J. 88:98–103.

Gettle, R.M., J.R. George, K.M. Blanchet, D.R. Buxton, and K.J. Moore. 1996b. Frost-seeding legumes into established switchgrass: Forage yield and botanical composition of the stratified canopy. Agron. J. 88:555–560.

Graber, L.F. 1927. Improvement of permanent bluegrass pastures with sweetclover. J. Am. Soc. Agron. 19:944–1006.

Hall, K.E., J.R. George, and R.R. Riedl. 1982. Herbage dry matter yields of switchgrass, big bluestem, and indiangrass with N fertilization. Agron. J. 74:47–51.

Heichel, G.H., and K.I. Henjum. 1991. Dinitrogen fixation, nitrogen transfer, and productivity of forage legume-grass communities. Crop Sci. 31:202–208.

Heichel, G.H., C.P. Vance, D.K. Barnes, and K.I. Henjum. 1985. Dinitrogen fixation, and N and dry matter distribution during 4 year stands of birdsfoot trefoil and red clover. Crop Sci. 25:101–105.

Heidemann, G.S., and G.E. Van Riper. 1967. Bud activity in the stem, crown, and rhizome tissue of switchgrass. J. Range Manage. 20:236–241.

Hunczak, J.J. 1992. Effects of early defoliation upon quality factors of switchgrass (*Panicum virgatum* L.). M.S. thesis. Iowa State Univ., Ames.

Jackobs, J.A. 1963. A measurement of the contributions of ten species to pasture mixtures. Agron. J. 55:127–131.

Jones, T.A., I.T. Carlson, and D.R. Buxton. 1988. Reed canarygrass binary mixtures with alfalfa and birdsfoot trefoil in comparison to monocultures. Agron. J. 80:49–55.

Jung, G.A., J.L. Griffin, R.E. Kocher, J.A. Shaffer, and C.F. Gross. 1985. Performance of switchgrass and bluestem cultivars mixed with cool-season species. Agron. J. 77:846–850.

Jung, G.A., J.A. Shaffer, and J.L. Rosenberger. 1991. Sward dynamics and herbage nutritional value of alfalfa-ryegrass mixtures. Agron. J. 83:786–794.

Mallarino, A.P., and W.F. Wedin. 1990. Effect of species and proportion of legume on herbage yield and nitrogen concentration of legume–grass mixtures. Grass For. Sci. 45:393–402.

Mallarino, A.P., W.F. Wedin, C.H. Perdomo, R.S. Goyenola, and C.P. West. 1990. Nitrogen transfer from white clover, red clover, and birdsfoot trefoil to associated grass. Agron. J. 82:790–795.

Marten, G.C. 1989. Summary of the trilateral workshop on persistence of forage legumes. p. 569–572. *In* G.C. Marten et al. (ed.) Persistence of forage legumes. ASA, CSSA, and SSSA, Madison, WI.

Marten G.C., W.F. Wedin, and W.F. Hueg, Jr. 1963. Density of alfalfa plants as a criterion for estimating productivity of an alfalfa–bromegrass mixture on fertile soil. Agron. J. 55:343–344.

Matches, A.G., and M. Mitchell. 1976. Growing legumes with warm-season grasses. Univ. of Missouri Spec. Rep. 192. Univ. of Missouri, Columbia.

Matches, A.G., and D. Mowery. 1979. Legumes frost-seeded into sods of switchgrass and caucasian bluestem. Univ. of Missouri Spec. Rep. 238. Univ. of Missouri, Columbia.

Mays, D.A., and E.M. Evans. 1972. Effects of variety, seeding rate, companion species, and cutting schedule on crownvetch yield. Agron. J. 64:283–285.

Nelson, C.J., and K.L. Larson. 1984. Seedling growth. p. 93–129. *In* M.B. Tesar (ed.) Physiological basis of crop growth and development. CSSA and ASA, Madison, WI.

Nichols, J.T., and J.R. Johnson. 1969. Range productivity as influenced by biennial sweetclover in western South Dakota. J. Range Manage. 22:342–347.

Olsen, F.J., J.H. Jones, and J.J. Patterson. 1981. Sod-seeding forage legumes in a tall fescue sward. Agron. J. 73:1032–1036.

Posler, G.L., A.W. Lenssen, and G.L. Fine. 1993. Forage yield, quality, compatibility, and persistence of warm-season grass–legume mixtures. Agron. J. 85:554–560.

Rountree, B.H., A.G. Matches, and F.A. Martz. 1974. Season too long for your grass pasture? Crops Soils 26(7):7–10.

Rowe, D.E. 1988. Alfalfa persistence and yield in high density stands. Crop Sci. 28:491–494.

Sharp, W.C., and G.A. Gates. 1986. Why use warm-season grasses in the Corn Belt and Northeast? p. 7–9. *In* Warm-season grasses: Balancing forage programs in the Northeast and Southern Corn Belt. Soil Conserv. Soc. of Am., Ankeny, IA.

Sims, P.L., L.J. Azuko, and D.N. Hyder. 1971. Developmental morphology of switchgrass and sideoats grama. J. Range Manage. 24:357–360.

Smith, D. 1979. Fertilization of switchgrass in the greenhouse with various levels of N and K. Agron. J. 71:149–150.

Smith, M. 1981. Switchgrass pastures in Iowa-varieties and N-fertilization. Iowa Coop. Ext. Serv. Pamph. CE-1602c. Iowa Coop. Ext. Serv., Ames.

Taylor, T.H., and L.T. Jones, Jr. 1983. Compatibility of switchgrass with three sod seeded legumes. p. 15. *In* Prog. Rep. of the Kentucky Agric. Exp. Stn. Univ. of Kentucky, Lexington.

Taylor, T.H., W.C. Templeton, Jr., and J.W. Wyles. 1973. Management effects on persistence and productivity of birdsfoot trefoil. Agron. J. 65:646–648.

Templeton, W.C., and T.H. Taylor. 1975. Performance of bigflower vetch seeded into bermudagrass and tall fescue swards. Agron. J. 67:709–712.

Waddington, J. 1992. A comparison of drills for direct seeding alfalfa into established grasslands. J. Range Manage. 45:483–487.

Weaver, J.E. 1968. Prairie plants and their environment. Univ. of Nebraska Press, Lincoln.

Wedin, W.F., J.D. Donker, and G.C. Marten. 1965. An evaluation of nitrogen fertilization in legume–grass and all-grass pasture. Agron. J. 57:185–188.

Wedin, W.F., and W. Fruehling. 1977. Smooth bromegrass vs. switchgrass for midsummer pasture. p. 9–11. *In* Ann. Prog. Report, Western Research Center, Castana, IA. ORC 77-10. Iowa State Univ., Ames.

Wedin, W.F., and W. Fruehling. 1980. Smooth bromegrass vs. switchgrass for mid-summer pasture. p. 9–11 *In* Ann. Prog. Rep., Western Research Center, Castana, IA. ORC 80-10. Iowa State Univ., Ames.

6   Improving Warm-Season Forage Grasses Using Selection, Breeding, and Biotechnology

Kenneth P. Vogel

USDA-ARS
Lincoln, Nebraska

Plant breeding is human-directed evolution. Plant breeders manipulate the genetic resources of a species, i.e., its germplasm, to produce plants that are of increased value to humanity. Although humans have successfully manipulated the genetic resources of plants and animals for several thousand years, the science of genetics was not developed until this century. Breeding work on most forage grasses in the USA did not began until the 1930s and initial work was focused on developing strains that had good establishment capability, persistence, high forage and seed yields, and good insect and disease resistance. These are essential attributes of forages (Burton, 1986). This initial breeding work resulted in the development of grasses such as 'Coastal' bermudagrass (*Cynodon dactylon* L.), 'Lincoln' smooth bromegrass (*Bromus inermis* Leyss.), and 'Kentucky 31' tall fescue (*Festuca arundinacea* Schreb.) (Vogel & Sleper, 1994). Limited animal evaluation was involved in the development of these cultivars. The initial breeding work on warm-season native grasses also began in the mid 1930s as a result of efforts to reseed land damaged by erosion, i.e., the dust bowl, in the Great Plains of the USA.

The native warm-season (C_4) grasses that have the most potential for use in agriculture are switchgrass [*Panicum virgatum* L.], big bluestem [*Andropogon gerardii* Vitman], and indiangrass [*Sorghastrum nutans* (L.) Nash]. They are tall warm-season (C_4) grasses that were the dominant grasses of the North American Tallgrass Prairie (Weaver, 1968). Although they are generally associated with the natural vegetation of the Great Plains and the Western Corn Belt, they occur widely throughout North America east of the Rocky Mountains and south of 55° N. Lat. in grasslands and nonforested areas (Hitchcock, 1971; Stubbendieck et al., 1991). They have been seeded in mixtures in the Great Plains for >50 yr as pasture and range grasses. In the past 20 yr they have become increasingly important as pasture grasses in the central and eastern USA because of their ability to be productive during the hot months of summer when cool-season grasses are relatively unproductive. Although the grasses belong to different genera, they have similar areas of adaptation, uses, and management requirements. The U.S. Department of Energy recently has identified switchgrass as a potential biomass fuel crop and is

---

Copyright © 2000. Crop Science Society of America and American Society of Agronomy, 677 S. Segoe Rd., Madison, WI 53711, USA. *Native Warm-Season Grasses: Research Trends and Issues.* CSSA Special Publication no. 30.

funding breeding and production research on this grass (Vogel, 1996). In a specific adaption zone switchgrass is generally the earliest of these grasses to mature and indiangrass is the latest (Moser & Vogel, 1995). Other warm-season grasses, both native and introduced, are being used in pasture and rangelands but big bluestem, switchgrass, and indiangrass have the broadest potential area of use.

REPRODUCTIVE AND BOTANICAL CHARACTERISTICS

Reproductive and botanical characteristics of a species determine the breeding procedures and breeding systems that can be used to develop improved cultivars and strains. The germplasm resources available to a breeder can limit the effectiveness of a breeding program. The two main germplasm resources for plants are native populations (in situ preservation) or seed of collections stored in germplasm depositories.

Big Bluestem

Big bluestem was the dominant species of the Tallgrass Prairie making up as much as 80% of the vegetation on some sites (Weaver, 1968). Today it can be found in remnant prairies or in old cemeteries throughout its former range of occurrence. These prairies and older varieties are the best germplasm sources of big bluestem available to plant breeders. Old varieties often represent the best germplasm from earlier collections. Databases of remnant prairies are maintained by state conservation agencies in cooperation with the Nature Conservancy. Seed of cultivars can be obtained from the releasing agency or from the U.S. Department of Agriculture's plant germplasm system. Sand bluestem is an ecotype or subspecies adapted to sandy soils such as the Nebraska Sandhills.

Big bluestem culms can grow to be 1 to 2 m tall and have numerous basal leaves. The vegetative portions of big bluestem plants are seldom over 50 to 60 cm high. Often the plants grow in clumps although most plants have short rhizomes (Stubbendieck et al., 1991). Sand bluestem has extensive rhizomes. Both big and sand bluestem can be propagated vegetatively by subdividing plants into clonal segments or ramets. In old plants, the crown can become dense and tough, almost woody, making dividing plants very difficult.

The inflorescence of big bluestem is a purplish-colored panicle with generally three digitate racemes, although it may vary from two to nine racemes. Another common name for big bluestem is Turkey Foot since the inflorescence resembles a turkey's foot. Spikelets are single-flowered and paired; the sessile spikelet is perfect and under most conditions, the pedicellata spikelet is staminate. Fertile pedicellata spikelets are found on some plants (Boe et al., 1983). The seed unit is the entire fertile, sessile spikelet that includes a rachis joint and the pedicel that supported the pedicellata spikelet. The seed has varying degrees of pubescence and has a twisted awn. These characteristics make unprocessed seed very fluffy and difficult to seed with conventional drills. Grassland drills have been developed that can plant the chaffy seed. The seed can be processed to remove the pubescence and appendages (Brown et al., 1983). Unprocessed seed weights typically average 550 seeds g^{-1} for big bluestem and 230 seeds g^{-1} for sand bluestem (Wheeler & Hill, 1957).

Big bluestem is cross-pollinated by wind and is largely self-incompatible (Law & Anderson, 1940; Norrmann et al., 1997). Flowering is triggered by long days followed by days with decreasing day length. Anthers begin shedding almost as soon as the inflorescence is out of the boot. Depending on strain and the latitude in which it is grown, pollen shed can occur from July to September. Pollen dispersal per plant may last 7 to 8 d because of varying maturity of tillers and florets (Jones and Newell, 1946). Less than 5% of pollen grains are dispersed more than 300 m (Jones & Newell, 1946). Pollen size is 35 to 50 µm in diameter (Jones & Newell, 1948). In the Sandhills region of Nebraska natural hybrids of big bluestem and sand bluestem are found on the bench areas above subirrigated meadows where big bluestem is found and the dune hills where sand bluestem is a dominant grass (Barnes, 1986). In controlled crosses, Peters and Newell (1961), demonstrated that sand and big bluestem were completely fertile with each other. Hybrids also were fertile.

The base chromosome number of big and sand bluestem is 10 (Gould, 1968). All released cultivars are $2n = 60$ (Riley & Vogel, 1982; Vogel, 1997 unpublished data); however, $2n$ chromosome numbers of 60 (hexaploids) and 90 (enneaploid) have been found in the same prairie location at several sites (Keeler 1990, 1992; Norrmann et al., 1997). The somatic nuclear DNA content is 5.93 pg and 8.92 pg, respectively, for the hexaploids and enneaploids (Norrmann et al., 1997). Meiosis in the hexaploids is largely normal with diploid pairing while meiosis in the enneaploid is irregular (Norrmann et al., 1997). The pairing behavior in the hexaploids at meiosis suggests an alloploid origin of the species (Norrmann et al., 1997). The enneaploids are believed to be produced from hexaploids by fertilization of an unreduced gamete by a normal gamete. In megasporogensis, an archesporial cell enlarges and undergoes meiosis to give a linear tetrad of megaspores (Norrmann et al., 1997). In hexaploids, the three megaspores close to the micropyle degenerate and the chalazal one, through successive mitotic division, develops into an embryo sac typical of sexually producing grasses (Norrmann et al., 1997). Microsporogenesis is usually normal in big bluestem (Dewald & Jahal, 1977). Consequently, quantitative genetic theory and breeding methods applicable to diploid plants can be applied to big bluestem.

Flowering occurs early in the morning with the peak time of pollen dispersal from 4 to 9 a.m. (central standard time; Jones & Newell, 1946; Norrmann et al., 1997). In controlled pollination studies, pollen germinates shortly after contacting the stigmata (Norrmann et al., 1997). In controlled self pollinations with hexaploids, growth of most pollen tubes is arrested at the stigma-style interface, <10% of the pollen tubes reach the ovaries and discharge their contents, and selfed seed set is usually <5% (Norrmann et al., 1997). In controlled pollinations between different hexaploid plants, pollen tubes grow readily through the stigma, style, and ovary, penetrate the micropyle within 2 h after pollination, and seed set is usually >50% (Norrmann et al., 1997).

Switchgrass

Switchgrass is an erect warm-season perennial grass (Hitchcock, 1971; Stubbendieck et al., 1991). It grows from 0.5 to 2 m tall and most tillers produce a seedhead when moisture is adequate. Although the plant can appear as a loose bunchgrass, it has short rhizomes and a stand has potential to thicken and form a sod. Plants can be easily subdivided into ramets. The inflorescence is a diffuse panicle 15 to 55 cm long with spikelets at the end of long branches. Spikelets are two-flowered with the second one being fertile and the first one sterile or staminate. The seed unit is a fertile floret. It is smooth and slick with an indurate lemma and palea that adheres tightly to the caryopsis (Wheeler & Hill, 1957). The seed threshes clean and is easy to process and plant. There are an average of 860 000 seeds kg^{-1} but there are large seed weight differences among cultivars (Wheeler & Hill, 1957). Johnson and Boe (1982) found 100-seed weights from 103 to 201 mg. Switchgrass is a cross-pollinated plant that is largely self-incompatible (Talbert et al., 1983). Switchgrass has a basic chromosome number of 9 and several levels of ploidy exist (Nielsen, 1944). Most switchgrass cultivars are either tetraploids or octaploids (Hopkins et al., 1996; Lu et al., 1998). Switchgrasses have been divided into lowland and upland types. Lowland types are taller, more coarse, generally more rust (*Puccinia graminis*) resistant, have a more bunch type growth and may be more rapid growing than upland types (Barnett & Carver, 1967; Brunken & Estes, 1975; Cornelius & Johnston, 1941; Porter, 1966). As indicated by the type description, lowland types are found on flood plains and other similar areas while upland types are found in upland areas that are not subject to flooding. Natural hybrids between the two types have not been identified to date. Because of the high degree of self-incompatibility in switchgrass, breeding methods that require use of inbred lines are not useful. Population improvement breeding methods applicable to diploid plants can be successfully used to improve switchgrass as long as separate tetraploid and octaploid populations are maintained.

Indiangrass

Indiangrass is a tall warm-season grass with short rhizomes. It has a loose bunch type growth habit since the rhizomes are generally shorter than 30 mm (McKendrick et al., 1975). Indiangrass height generally ranges from 0.5 to 2 m and has a yellowish brown to black panicle that ranges from 10 to 30 cm in length (Hitchcock, 1971; Stubbendieck et al., 1991). It belongs to the Andropogoneae tribe, as does big bluestem, and the spikelet and floret structure is similar to big bluestem. The spikelets are in pairs on the rachis with the sessile one fertile and the pedicellata one rudimentary or absent (Stubbendieck et al., 1991). As with big bluestem, the spikelets disarticulate below the glumes, the glumes and florets are covered with pubescence, and the fertile floret has a twisted awn. The seed unit is the spikelet with appendages and it is extremely fluffy and difficult to plant with conventional grass drills unless it is processed to remove pubescence and seed appendages. For unprocessed seed, there are about 385 seeds g^{-1} while seed weight of caryopses ranges from 120 to 150 mg 100 seed^{-1} (Wheeler & Hill, 1957; Barnett & Vanderlip, 1969). Some indiangrass seedlots have a considerable amount of dormancy (Emal & Conard, 1973), depending on cultivar and the season of seed production.

A major portion of indiangrass tillers have been described as biennial (McKendrick et al., 1975). The first year they grow vegetatively and then after overwintering, a tiller will become reproductive. This characteristic can be easily seen by observation of the earliest growth in the spring. The tillers, however, do not appear to be obligate biennials since indiangrass will flower the seeding year. Indiangrass normally may flower about 4 to 6 wk later than switchgrass collected in the same area (McKendrick et al., 1975). As the genus name *Sorghastrum* implies, indiangrass morphologically and chemotaxonomically appears to be closely related to the sorghums *(Sorghum* sp.). Indiangrass is the only forage grass outside of the sorghum genus known to contain cyanogenic glucosides (Gorz et al., 1979). Several groups have attempted to make sorghum–indiangrass hybrids but with no success (Pedersen et al., 1993). Recent comparisons between indiangrass and sorghum using chloroplast restriction fragment length polymorphisms (RFLPs) indicates that the two species are not closely related (Pedersen et al., 1993).

All of the published cytogenetic reports on North American indiangrasses indicate that they have $2n = 40$ chromosomes and that meiosis is normal (Bragg, 1964; Church, 1929; Riley & Vogel, 1982). Gould (1975) reported indiangrass to have a complement of $2n = 20$, 40, and 80, although he later indicated that all counts in his records were $2n = 40$ except for one count of $2n = 20$ from Brazil (Gould, 1979, personal communication). The cultivars Nebraska 54, Oto, Holt, and Osage have chromosome numbers of $2n = 40$ (Riley & Vogel 1982). Indiangrass is cross-pollinated by wind. Some plants will set considerable amounts of selfed seed when several panicles are bagged together in parchment bags but other plants produce almost no seed when selfed (Newell, 1936, unpublished data). In the same reports, Newell noted that selfed lines were similar to open-pollinated lines in germination and establishment. Most of the pollen is shed during an 8-d period with peak times from 6 to 10 a.m. (Jones & Newell 1946). Because of its reproductive behavior, breeding methods that can be used on big bluestem and switchgrass also can be used on indiangrass.

GENETIC VARIATION

Plants of switchgrass, big bluestem, and indiangrass are photoperiod sensitive that is under genetic control. Photoperiod requirements for flowering and fall senescence differ depending on latitude of origin and are the primary factors determining cultivar adaptation (Moser & Vogel, 1995). Southern strains moved north will be exposed to longer than normal photoperiods for a given date because of latitude and as a result they will stay vegetative longer and produce more forage (Cornelius & Johnston, 1941; McMillian, 1959, 1965; Newell, 1968). Northern strains moved south will be exposed to a shorter than normal photoperiod and will flower early. In a controlled study, Benedict (1941) determined that switchgrass requires short days to induce flowering. In nature, flowering is induced by decreasing day length during early summer. The photoperiod response also appears to influence winter survival. Southern types moved too far north will not survive winters because they stay vegetative too late in the fall. As a general rule, these grasses should not be moved more than 500 km north of their area of origin because of the possibility of stand losses due to winter injury. The response to photoperiod can be modified by growing degree days to some extent since the flowering date of cultivars will vary from year to year. In addition to photoperiod, the other factor that determines specific adaptation is response to precipitation and the associated humidity (Moser & Vogel, 1995). Cultivars from the more arid Great Plains states may develop foliar disease problems when grown in the more humid eastern states. Cultivars based on eastern germplasm may not be as well adapted to drought stress as cultivars based on western germplasm.

Although much of the prairie and grasslands that were once occupied by these grasses were plowed and converted into cropland, remnant prairie sites still exist in most areas and are an invaluable germplasm resource. Genetic variation exists among germplasm collected from specific regions (ecotype variation), among accessions of an ecotype (population variation), and among plants of a population collected from a specific site (Eberhardt & Newell, 1959; Hopkins et al., 1995b; Vogel & Pedersen, 1993; Vogel et al., 1994; Vybiral et al., 1993). The natural genetic variation that can be used by plant breeders to improve these grasses is the ecotype and among and within population genetic variation that exists within these species. Genetic studies that have been completed on these grasses are listed below.

Switchgrass: Anderson et al., 1988; Boe & Johnson, 1987; Eberhardt & Newell, 1959; Gabrielsen et al., 1990; Godshalk et al., 1986, 1988a,b; Godshalk & Timothy, 1988a,b; Gunter et al., 1996; Hopkins et al., 1993, 1995a,b, 1996; Hultquist et al., 1996, 1997; Johnson & Boe, 1982; Newell & Eberhardt, 1961; Talbert et al., 1983; Vogel et al.,1981c, 1984, 1993.

Big and sand bluestem: Barnes, 1986; Boe & Ross, 1983; Boe et al.,1983, 1989; Glewen & Vogel, 1984; Kneebone, 1956; Newell, 1968; Newell & Peters, 1961; Peters & Newell, 1961; Ross et al., 1975; Riley, 1981.

Indiangrass: Barnett et al., 1971; Barnett & Vanderlip, 1969; Kube et al., 1989; Vogel et al., 1980, 1981a,b, 1994; Weimer et al., 1988.

The conclusions from these studies can be summarized as follows: (i) there is substantial genetic variability both between and within strains of these grasses for most agronomic traits including those that affect or determine forage yield, seed yield, and forage quality and; (ii) heritability values for most important traits range from 20 to 40% that should make it possible to improve these grasses by breeding; and (iii) correlations among most desirable traits are usually positive, but when negative they are usually not large indicating it should be possible to simultaneously improve several traits without adversely affecting other traits. Molecular genetic research with RFLP and RAPD markers (Hultquist et al., 1996, 1997; Gunter et al., 1996) demonstrates that molecular genetics techniques can be valuable in breeding programs for classifying germplasms. Flow cytometry analyses for determining ploidy level of individual plants for assignment to specific breeding populations is developing into a very effective research tool for grass breeding programs (Hultquist et al., 1996, 1997; Hopkins et al., 1996).

BREEDING SYSTEMS

The improvement that can be made in a plant breeding program is dependent upon the genetic variability within the species for the traits being selected, the heritability of the traits, the breeder's ability to identify genetically superior plants, the intensity of selection, and the effectiveness of the breeding procedure (Allard,

1964; Hallauer & Miranda, 1981). Existence of genetic variability is essential because without it, no progress can be made by breeding. Heritability is the proportion of the variability for a particular trait in a population that is due to genetic differences among plants. Traits such as forage yields in grasses are determined by many genes and are greatly influenced by environment and hence have low heritability values.

Ecotypes or endemic strains found in specific regions and sites have evolved by the genetic mechanisms of mutation, migration, selection, and random drift or chance resulting in between ecotype or endemic strain genetic variability (Falconer, 1981). Eberhardt and Newell (1959) and Hopkins et al.(1995b) documented this between strain variability in switchgrass. Most of the initial breeding work with cross-pollinated grasses used between strain genetic variability (Vogel & Gabrielsen, 1986; Vogel & Pedersen, 1993; Alderson & Sharp, 1994). A large array of native accessions (ecotypes or strains) were collected from the general geographical area (native grasses) or from a climatic analogue (introduced grasses) in which a cultivar of the grass was needed (Fig. 6–1). These collections then were evaluated in uniform nurseries for various agronomic traits. The better accessions were increased for testing in additional environments and based on these tests released directly to the public without any additional breeding work. This procedure was used by many of the state experiment stations and the Plant Materials Centers of the USDA-SCS in developing the initial cross-pollinated grass varieties for different geographical regions of the USA (Alderson & Sharp, 1994). Examples of grass cultivars developed by this procedure include Blackwell and Nebraska 28 switchgrass. Collection and evaluation of plant introductions and native accessions is an ongoing process with a high national priority since the superior strains identified by this process form the germplasm base for the further improvement of grasses by breeding.

Within strain genetic variability consists of the proportion of the plant-to-plant (phenotypic) variability that exists between plants of a strain that is due to genetic (genotypic) differences among plants (Falconer, 1981; Hallauer & Miranda, 1981). This variability is very difficult to observe or measure in a typical pasture or rangeland situation. If, however, seed is harvested from the individual plants in a common native prairie site and planted in a space-planted nursery under uniform conditions, phenotypic variation among plants can be readily distinguished. By using population or quantitative genetics procedures, geneticists can determine the proportion of the total plant-to-plant phenotypic variability that is due to genetic differences among plants, i.e., they determine the total genetic variability in a population for specific traits and the heritability of those traits.

Switchgrass, big bluestem, and indiangrass have small florets that are difficult to emasculate, and effective mechanisms for producing hybrids such as cytoplasmic male sterility have not been developed for these grasses. Thus, breeders are limited largely to procedures that use additive genetic variability and that do not require emasculation (Vogel & Pedersen, 1993). Fortunately, there is substantial additive genetic variability for most traits in these grasses, and efficient breeding methods that do not require emasculation are available for perennial plants that can be vegetatively propagated. Breeding schemes or methods that can be used on perennial grasses such as big bluestem, indiangrass, and switchgrass and the expected gain from selection that can be made by using the various schemes are described by Vogel and Pedersen (1993). The two breeding methods that have the most potential of exploiting additive genetic variation are Restricted Recurrent Phenotypic Selection or RRPS (Burton, 1974, 1982) and a modified form of between and within half-sib family selection (B&WFS) (Asstveit & Asstveit, 1990; Vogel & Pedersen, 1993). Both methods are population improvement breeding procedures that use recurrent selection and take advantage of the perennial nature of these grasses and their ability to be vegetatively propagated.

The objective of a recurrent selection breeding program is to change the frequency of desirable genes in a the population and by doing so change the popula-

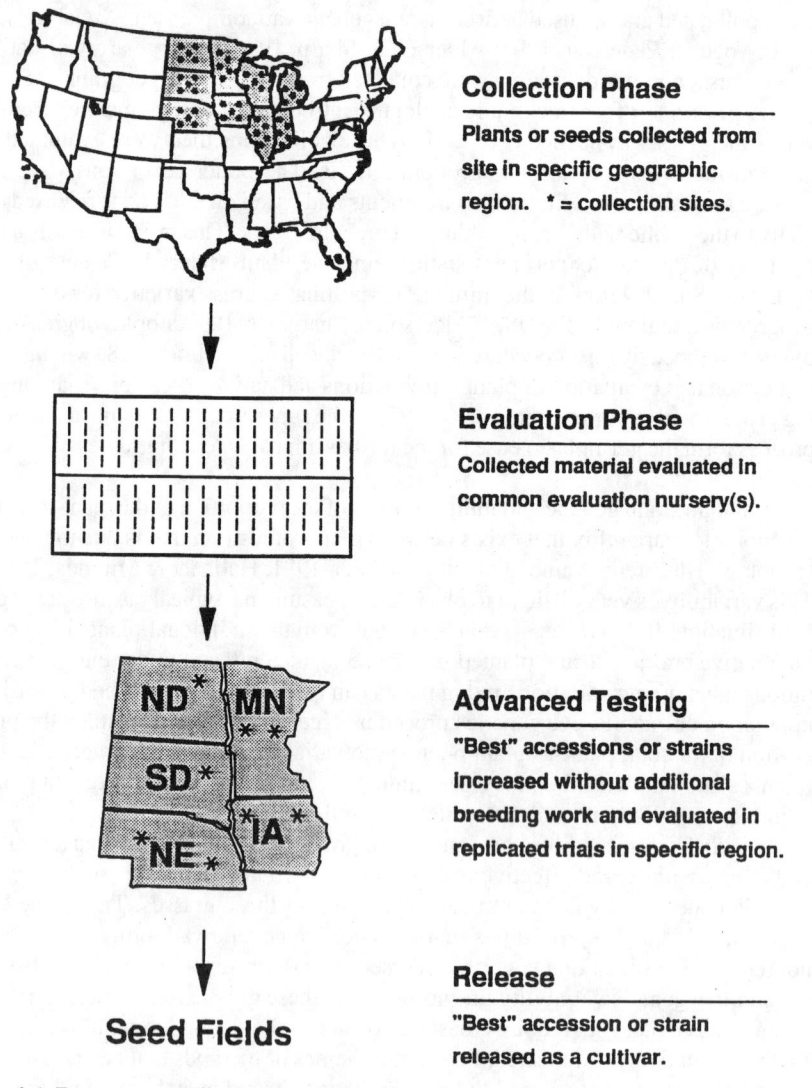

Collection Phase
Plants or seeds collected from site in specific geographic region. * = collection sites.

Evaluation Phase
Collected material evaluated in common evaluation nursery(s).

Advanced Testing
"Best" accessions or strains increased without additional breeding work and evaluated in replicated trials in specific region.

Release
"Best" accession or strain released as a cultivar.

Seed Fields

Fig. 6–1. Ecotype or naturalized strain selection. Sites of seed collection are indicted by asterisks.

tion mean. In a theoretical example (Fig. 6–2), the relative yield of a population was improved 16% in three breeding cycles.

In both RRPS and B&WFS breeding systems (Fig. 6–3 and 6–4), greenhouse grown seedlings are transplanted into field evaluation nurseries. Limited data are collected the establishment year. The following years are used to collect yield, forage quality data, and other agronomic information. This data then is used to select superior plants that are moved into polycross nurseries. Seed harvested from the

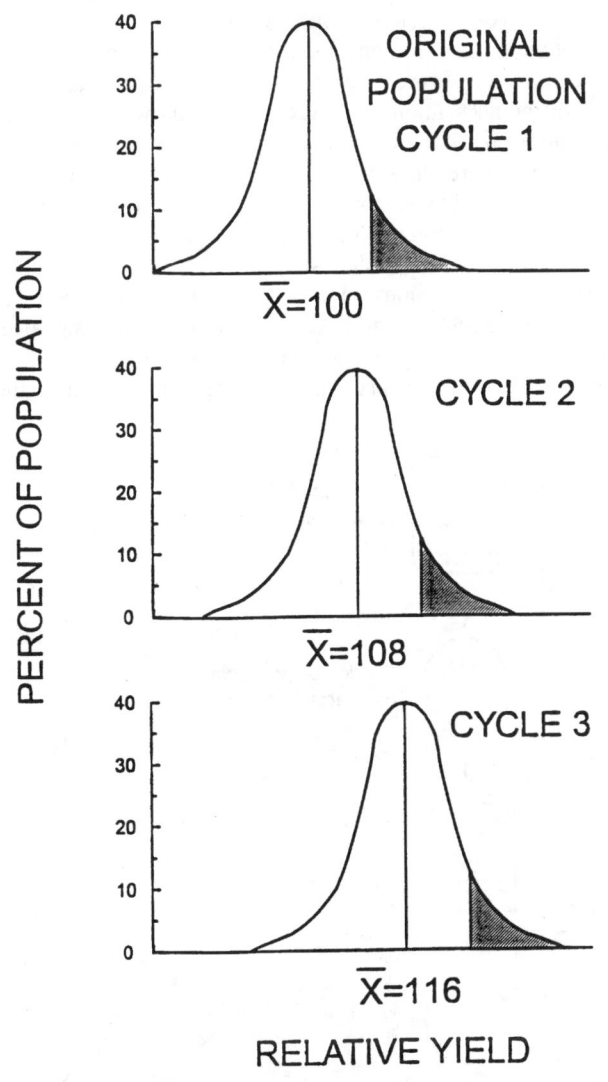

Fig. 6–2. Representation of the theoretical effect of three cycles of restricted, recurrent phenotypic selection on yield. The area under the curve represents all the plants in the population. The shaded area represents the selected plants. In this example 5% of the highest yielding plants are selected from each cycle, heritability is 40% and the phenotypic standard deviation is 10. The population mean ($\bar{x}$) of the base population (C0) is 100 in Cycle 1.

polycross nurseries is used to start the next cycle of selection and for yield trials. Each breeding cycle will require 4 or 5 yr. A four year cycle is: Year 1—establishment; Year 2 and 3—collect data and make selections; Year 4—polycross. A 4-yr cycle is being used because studies currently being completed indicate that in switchgrass, selection for high in vitro dry matter digestibility (IVDMD) can affect winter survival in some populations and several seasons are necessary to eliminate nonwinter hardy plants. A 5-yr cycle could be used in B&WFS if 2 yr are used to evaluate families and then 1 yr is used to identify superior individuals in the best families. RRPS was very effective in improving in vitro dry matter digestibility in switchgrass, but yield improvements have been small (Hopkins et al., 1993). The between and within family breeding procedure (Fig. 6–4) provides a method of measuring genetic differences among families and also allows the breeder to maintain adequate population size to avoid inbreeding depression. The expected gains from selection for these two breeding methods are similar if the breeder uses the same number of years per cycle (Vogel & Pedersen, 1993). The time period required to develop and test a new cultivar can exceed 15 yr particularly if grazing trials are part of the evaluation process (Table 6–1).

Direct selection for a single trait usually will result in the maximum gain from selection for that trait. Multiple trait selection adds to the challenge because the desired traits may have low or negative genetic correlations with one another. In general, grass breeders have made very limited use of formal selection indexes based

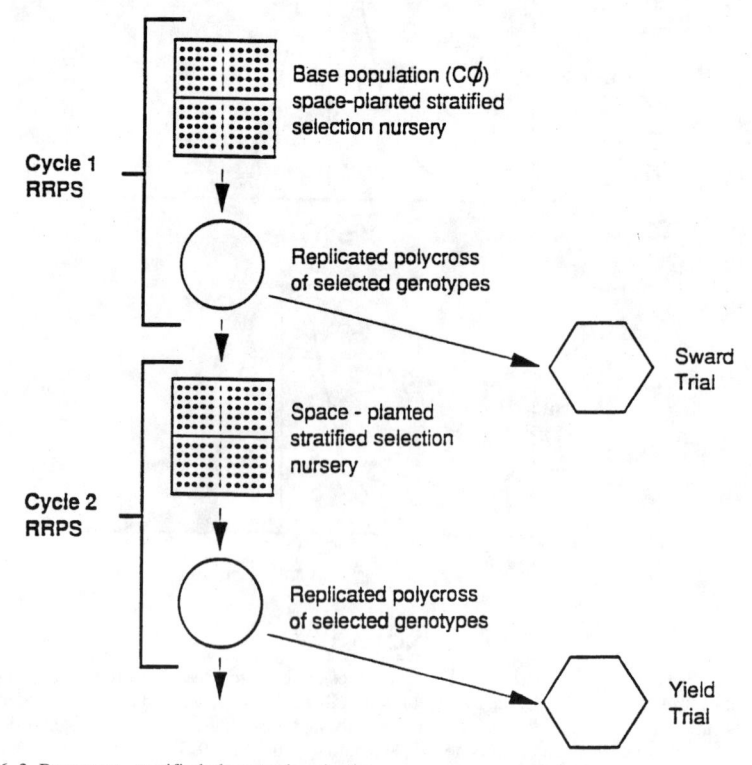

Fig. 6–3. Recurrent, stratified phenotypic selection.

on quantitative genetic theory but have instead relied on informal subjective indexes. Research by Godshalk and Timothy (1988a,b) indicate that selection indexes may be effective in breeding for multiple traits in perennial grasses. The use of selection indexes based on quantitative theory in which traits are weighted by realistic economic values should result in improved breeding efficiency.

Grass breeders in general have not capitalized on the nonadditive genetic variability that exists in forage grasses even though substantial heterosis for traits such as yield exists in many grasses. The inability to effectively emasculate large numbers of plants in seed production fields has limited grass breeders' ability to develop hybrids for commercial use. Techniques that may have application for big bluestem, switchgrass, and indiangrass are: production of first generation chance hybrids, hybrids produced by using self-incompatibility, and the use of male gametocides (Burton, 1986; Vogel et al., 1989). First generation chance hybrids are produced by mixing seed of four lines or families, planting the bulk, and then harvesting the seed. The seed harvested from the seed field should contain 75% hybrid seed of the six possible hybrids (Burton, 1948). Alternatively, if two plants are identified that are self-sterile but cross fertile and produce superior F_1 hybrids, then the two plants can be vegetatively increased and transplanted into seed production fields. All the seed harvested from the field would be F_1 hybrid seed assuming that proper isolation requirements were maintained. These seed fields could be maintained for many

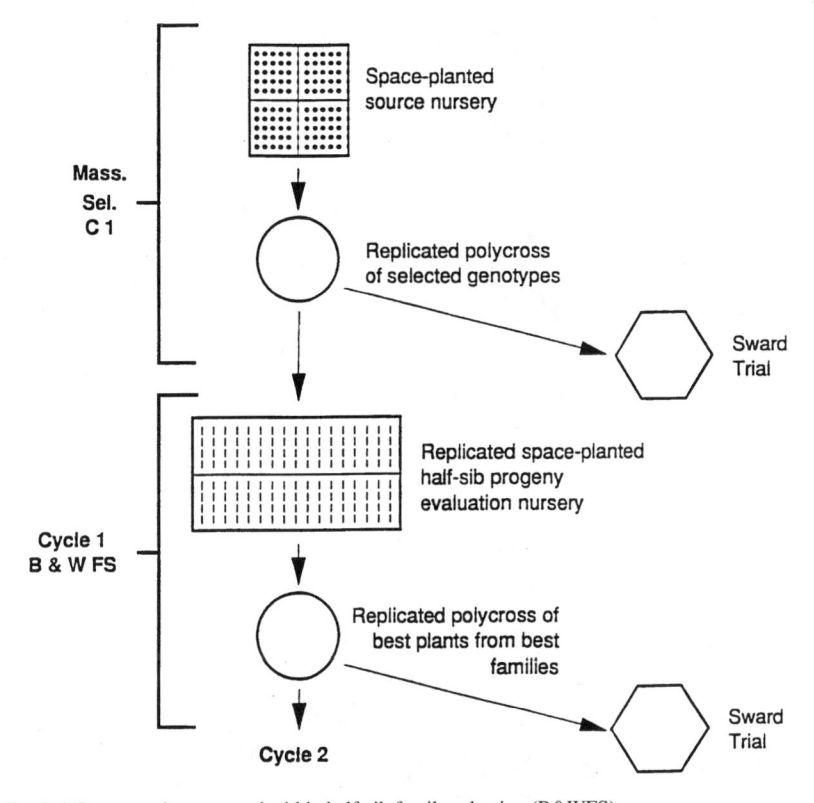

Fig. 6–4. Recurrent between and within half-sib family selection (B&WFS).

Table 6–1. Project timetable for a forage grass breeding program.

Phase	Year 1	Year 2	Year 3	Year 4	Year 5
1. Germplasm evaluation and selection nurseries	Establish native grass evaluation studies or selection nurseries	Evaluate forage yields, quality, and other traits	Second year of evaluation	Identify superior plants and move to crossing blocks, initial seed harvest	Harvest seed, repeat cycle in breeding program. Use seed to plant regional trials, or release as germplasm
2. Regional small plot trials	Plant trials	Harvest trials	Harvest trials	Summarize data, begin seed increase of best strains for pasture trials or release	Seed harvested from increase nurseries
3. Grazing trials of advanced lines	Plant pastures	Grazing trial	Grazing trial	Increase best strain for release/ continue grazing to monitor persistence	Release seed to seed growers

years. This method of producing hybrids is currently being evaluated at Lincoln, NE. A method that has potential for producing hybrids is the use of male-gametocides in which a chemical is used to emasculate the female strain. Since many of these gametocides are proprietary compounds, the necessary research would have to be done in conjunction with firms that have ownership of the compounds.

Thus far, grass breeders have used conventional breeding techniques to develop new cultivars. New technologies are becoming available for breeders to greatly expand their capabilities to solve specific breeding problems. These technologies, which are referred to as cell culture and molecular genetics, must be used in conjunction with conventional breeding methods because their sole use would not result in the development and use of commercial cultivars. Techniques to culture individual plant cells and to regenerate plants from these cells have been developed for big bluestem, indiangrass, and switchgrass (Chen & Boe, 1988; Denchev & Conger, 1994). Tissue culture gives breeders the capability to rapidly and efficiently multiply individual plants that could make commercial self-incompatible hybrids feasible if somaclonal variation can be controlled. A new method for producing clonal propagules of switchgrass without somaclonal variation has recently been developed (Alexandrova et al., 1996).

Tissue culture also enables breeders to select and apply mutagenic treatments at the cellular level or to genetically alter plants via genetic transformation. A very effective transformation method involving particle bombardment is available for wheat (Weeks et al., 1993). Dr. Bob Conger has successfully transformed switchgrass using particle bombardment as indicated by marker genes (personal communication, 1997). The technology is available to develop effective transformation procedures for any monocot species if adequate financial, laboratory, and personnel resources are available. Molecular genetics techniques as applied to plant breeding will be used primarily to transfer traits to plants from very dissimilar organisms or to block expression of specific genes. Molecular genetics techniques permit the transfer of specific genes rather than whole blocks of genes. Traits that are being transferred into monocots include genes for insect, disease, and herbicide resistance and for unique proteins or other plant products. The main limitation for use of transformation procedures for grasses are the limited financial resources available for use on specific species and the high costs involved in getting transformed plants approved for use in production systems. Most forage programs simply do not have the resources to get transformed perennial grasses cleared for use in production agriculture. Because of the potential gene flow from cultivars to native populations of warm-season grasses, obtaining clearance to use transformed plants of native grasses in production agriculture may be more difficult than obtaining clearance for crops such as maize (*Zea mays* L.).

The objectives of most forage grass breeding programs are to improve one or more of the following traits: establishment capability, persistence, forage yield, forage quality, seed yield, and disease and insect resistance (Vogel et al., 1989).

Improved Establishment

Rapid and reliable establishment is critical for the conversion of cropland to grassland. Seedling establishment can be improved by either improving the establishment capability of the plants by breeding or by modifying the environment with cultural practices. The primary causes of poor establishment are the related factors of moisture stress and weed competition. It is usually not feasible in the Great Plains to modify the environment by irrigating for grassland establishment.

Breeding for improved seedling vigor can result in seedlings that develop rapidly and that are effective competitors with weeds for available moisture and sunlight. Factors that affect seedling vigor are seed size, seed quality, germination rate, emergence rate, relative growth rates and other physiological processes (McKell, 1972). In virtually all the studies that have been done to date, seed size or weight has been an important component of seedling establishment capability (Asay & Johnson, 1987; Voight et al.,1987, Kneebone, 1956). Breeding for improved seed size can result in strains with improved seedling growth and establishment (Trupp & Carlson, 1971; Wright, 1977; Clements and Latter, 1973).

Seed size, however, is only one component of seedling growth. Whalley et al. (1966) subdivided seedling growth into a heterotrophic, transition, and autotrophic stages. The heterotrophic stage included all physiological activities prior to the initiation of photosynthesis, and the autotrophic stage occurs after all endosperm reserves have been depleted. The effect of seed size probably occurs in the heterotrophic and transitional stages and it may have a carryover effect in the autotrophic stage. Variation for seedling growth within a species therefore is due to variation for seed weight and seedling vigor. In sand bluestem, 50% of the genetic variability for seedling weight 8 wk after establishment was due to seed size differences while the remainder was due to other factors (Glewen & Vogel, 1984). Seedling vigor is difficult to measure in grasses because it can be affected by growing conditions in seed production fields, seed harvesting, handling, storage procedures, and by varying degrees of seed dormancy when seeds are planted in soils varying in temperature.

Two procedures recently have been used to successfully breed for improved seedling vigor in native, warm-season grass. I (Sebolai & Vogel, 1989) used stratified mass selection to select for individual seedling weight of big bluestem, switchgrass, and indiangrass grass seedlings grown in individual micro-pots (cone-tainer cells). Seedlings were cut and weighed 6 wk after planting, the heaviest seedlings were allowed to regrow, and then were transplanted into field polycross nurseries. Three cycles of selection have been competed in switchgrass, big bluestem, and indiangrass populations. Evaluation trials have indicated that significant improvements in seedling vigor have been made using this procedure (Sebolai Vogel, 1989). Boe and Johnson (1987) conducted mass selection for seed size in a switchgrass population by separating heavy seed from light seed with a seed blower and significantly improved seed size. Selection for seedling weight or seed size could be used in tandem with selection for other traits such as forage quality.

Persistence

Developing cultivars from germplasm indigenous to the region of use or from an area of the world with similar climate and soils is the primary procedure that is used to breed for persistence (Vogel et al., 1989; Hanson & Carnahan, 1956). Persistence problems with native warm-season grasses can be solved by using cultivars that originate from the same Plant Hardiness Zone in which they are used in production agriculture. Persistence problems develop when cultivars are moved too far north or south from their latitude of origin. Because germplasm resources for switchgrass, big bluestem, and indiangrass are available for most areas where they will be used, it usually is easier to use adapted material in breeding programs rather than attempting to change the adaptability of a strain by breeding.

Breeding work to improve disease and insect resistance can improve persistence; however, sometimes breeding for specific traits can affect persistence adversely. Recently completed research on a switchgrass population that has gone through three cycles of selection for high IVDMD has indicated that breeding work to increase IVDMD has produced a population with reduced ability to survive Nebraska winters (Vogel, 1991–1995, unpublished data). Some families in the population, however, have good winter survival indicating that it should be possible to continue to improve IVDMD but attention will have to be given to persistence. Bahiagrasses selected for high yield in a rapid generation RRPS program also showed reduced winter survival in Florida (Pedreira & Brown, 1996). Although rapid generation turnover theoretically can maximize gain from selection, these results indicate that perennials expected to persist in production pastures for 10 yr or longer may require longer evaluation periods in breeding nurseries.

Seed Yields

Initially, seed production was a problem with most perennial grasses adapted to the Great Plains. Empirical (Schumacher, 1962) and formal research (Cornelius, 1950; Kassel et al., 1985) has resulted in greatly increased seed production. By using improved seed production practices, experienced seed growers in Nebraska can produce from 250 to >1000 kg of seed ha^{-1} (Nebraska Crop Improvement Association, 1990, personal communications). Since only 6 to 14 kg of pure live seed ha^{-1} are needed for grassland plantings, one seed production hectare will plant 25 to 90 ha of grasslands, which is similar to that of most cultivated crops (Vogel et al., 1989). Breeding for seed yield for species where seed yields are adequate and the end product is biomass thus does not appear to be warranted.

The low seed yields that occur periodically in the region probably are due to insect predation. A big bluestem seed midge (*Contarinia wattsi* Gagné) can reduce yields of these grasses by >50% (Carter et al., 1988). Limited information on the biology of the bluestem seed midge is available (Vogel & Manglitz, 1989). Systemic insecticides may be effective but two that were evaluated, carbofuran and orthene, were not effective in reducing seed losses (Vogel & Manglitz, 1990). Based on existing knowledge of the insect, it is likely that light applications of insecticides timed to match the emergence of the adult could be an effective control. Evaluating non-systemic insecticides will require multiple isolated fields in which treatments can be applied. Boe et al. (1989) recently reported evidence that there may be genetic variation for infestation by the big bluestem seed midge. The genetic differences were small and a tremendous amount of long-term breeding work (20+ yr) would be required to produce strains that had economically improved tolerance or resistance. A limited amount of entomological work on finding a suitable insecticide would be a better use of research funds. Seed midges also have been reared from midge pupae found in indiangrass and little bluestem (Vogel & Manglitz, 1989).

Disease and Insect Resistance

Grass breeders have made significant improvements in disease tolerance and resistance of forage grasses including native warm-season grasses. Virtually every improved cultivar on the market today is superior in disease resistance to common strains or native collections. Additional genetic gains in disease resistance can be made by continued selection for plants free from diseases in breeding programs. Genetic sources of resistance have been reported for almost every disease of economically important cool-season grasses (Braverman, 1986). A similar situation probably exists in warm-season grasses. A problem with breeding for disease resistance in warm-season grasses has been the lack of basic plant pathology work on disease characterization and identification. Research by Zeiders (1982, 1984) and Snetskaar and Tiffany (1990) on diseases affecting warm-season grasses has been very helpful.

Forage Yield

Improving forage yield always has been one of the principal objectives of grass breeders. The easiest way to breed for improved yield in native warm-season grasses is to move southern strains north. There is a problem associated with this approach since strains moved too far north will winter kill. For example, switchgrass strains from southern Oklahoma and Texas do not survive Nebraska winters. Recent reports document that forage yields can be improved significantly by breeding with substantial economic benefits (Vogel et al., 1989). Burton (1982, 1985) improved forage yield of bahiagrass (*Paspalum notatum* Fluegge) by direct selection for yield using RRPS. Nelson et al. (1985) improved forage yield of tall fescue (*Festuca arundinaceae* Schreber) by selection for leaf area expansion rate. These breeding efforts were successful because the breeders either selected directly for yield or for a trait that was correlated with yield, and they used recurrent selection methods that effectively exploited the additive genetic variability for the selected traits within the species. It should be possible to improve yield of most forage grasses by using well-designed recurrent selection methods. Breeding for yield remains a valid research objective.

Forage Quality

An extensive review of breeding for forage quality was completed recently by Vogel and Sleper (1994). Forage quality can be improved by breeding for enhanced positive quality factors such as digestibility or for reduced negative factors such as alkaloids (Burton, 1981; Marten, 1989). There are no reported problems in big bluestem with any anti-quality factors such as alkaloids or endophytes. Pouli et al. (1992) reported that a photo-sensitization factor is present in switchgrass that affects sheep. In the Central Great Plains and the Midwestern states, switchgrass is largely used by beef cattle so the reported photo-sensitization factor has not been a problem in these regions. Indiangrass contains dhurrin (a cyanogenic compound) that can cause prussic acid poisoning of livestock (Gorz et al., 1979). The hydrocyanic acid potential (HCN-p) of indiangrass exceeds critical levels only in new growth of pure stands of indiangrass that is <20 cm tall (Vogel et al., 1987). Indiangrass in mixed stands in seed pastures would be diluted by other grasses and forbs so the potential for hydrocyanic poisoning of livestock is reduced. No deaths of livestock due to grazing indiangrass have been reported.

Emphasis in breeding programs can be placed on improving positive quality factors because of the limited anti-quality problems of big bluestem, switchgrass, and indiangrass. Significant gains have been made in improving digestibility of switchgrass that has resulted in improved animal performance (Anderson et al., 1988; Vogel et al., 1993). Increased quality results in increased net return to a livestock producer and does not require any additional investment. Increased yield can increase net return, but the producer must buy or raise additional livestock to use the additional forage. Digestibility of forage is primarily determined by the plant cell walls that are composed of cellulose, hemicellulose, and lignin. Factors affecting cell wall digestion are complex and are related to lignin concentration and composition. Cell wall composition including lignin composition is under genetic control. Fortunately, the in vitro dry matter digestibility test measures the cumulative effects of the various factors and gives forage breeders a single value upon which to make selections. Genetic variation exists in switchgrass, indiangrass, and big and sand bluestem for forage digestibility (Godshalk et al., 1986; Riley, 1981; Ross et al., 1975, Vogel et al., 1981a,b) so genetic gains to improve IVDMD of these grasses can be achieved. The development of instruments, associated software, and procedures for such methodology as near infrared reflectance spectroscopy (NIRS), high performance liquid chromatography (HPLC), gas chromatography, and other procedures has greatly expanded the capability of grass breeders to select for specific plant composition and constituents.

Currently research is in progress to modify plant cell walls by using molecular genetics technology to block or alter specific steps in the lignin synthesis pathway using anti-sense technology. Transgenic tobacco with altered cell walls has been produced (Halpin et al., 1994). This technology can be transferred to monocots and will probably be applied to annual monocot crops such as maize grown for silage. Using this technology on perennial, polyploid grasses will be more difficult because the affect of specific pathways on the persistence of perennials is unknown. Even after transformed plants are developed, the expense of clearing transformed plants for use in production agriculture will exceed the capability of most forage breeding programs. Because significant levels of natural variation exist in most forage grasses for forage quality traits (Vogel & Sleper, 1994), this source of genetic variation should be exploited before more expensive genetic transformation procedures are used.

CULTIVARS

Cultivars of big bluestem, switchgrass, and indiangrass have been developed and released by state experiment stations, by USDA-ARS breeding programs, and by the Plant Material Centers of USDA-NRCS. The principal cultivars that are available for use as forage grasses are summarized in Tables 6–2, 6–3, and 6–4 for switchgrass, big and sand bluestem, and indiangrass, respectively. The adaptation zone for cultivars can be based on the USDA Plant Hardiness Zone Map (Fig. 6–5) since

Table 6–2. Principal cultivars of switchgrass and their areas of adaptation. Origin is area where germplasm for cultivars was collected. Adaptation zones are indicated on USDA Plant Hardiness Zone Map (Fig.6–5).

Cultivar	Origin	Type	Chromosome number ($2n$)	Adaptation zone
Dacotah	North Dakota	upland	36	2, 3, upper 4
Forestburg	South Dakota	upland	72	3, 4
Sunburst	South Dakota	upland	72	3, 4
Nebraska 28	Nebraska	upland	72	3, 4
Summer	Nebraska	upland	36	3, 4, 5
Pathfinder	Nebraska, Kansas	upland	72	4, 5
Trailblazer	Nebraska, Kansas	upland	72	4, 5
Blackwell	Oklahoma	upland	72	lower 5, 6, 7
Cave-in-Rock	Southern Illinois	upland	72	5, 6, 7
Shawnee	Southern Illinois	upland	72	5, 6
Shelter	West Virginia	upland	72	4, 5
Kanlow	Oklahoma	lowland	36	6, 7
Alamo	Texas	lowland	36	7, 8, 9

Table 6–3. Principal cultivars of big and sand bluestem and their areas of adaptation. Origin is area where germplasm for cultivars was collected. Adaptation zones are indicated on USDA Plant Hardiness Zone Map (Fig. 6–5).

Cultivar	Origin	Type	Chromosome number ($2n$)	Adaptation zone
Bison	North Dakota	big bluestem	60	2, 3
Bonilla	South Dakota	big bluestem	60	3, upper 4
Niagara	New York	big bluestem	60	3, 4, upper 5
Champ	Nebr., Iowa	intermediate	60	4
Pawnee	Nebraska	big bluestem	60	lower 4, 5
Rountree	Iowa	big bluestem	60	4, 5, upper 6
Garden	Nebraska	sand bluestem		4, 5
Goldstrike	Nebraska	sand bluestem	60	4, 5
Kaw	Kansas	big bluestem	60	lower 5, 6, 7
Earl	Texas	big bluestem	60	7, 8

Table 6–4. Principal cultivars of indiangrass and their areas of adaptation. Origin is area where germplasm for cultivars was collected. Adaptation zones are indicated on USDA Plant Hardiness Zone Map (Fig. 6–5).

Cultivar	Origin	Chromosome number ($2n$)	Adaptation zone
Tomahawk	North & South Dakota		3, upper 4
Holt	Nebraska	40	4
Nebraska 54	Nebraska	40	lower 4, 5
Oto	Nebraska, Kansas		5, upper 6
Rumsey	Illinois		5, 6
Osage	Kansas, Oklahoma	40	6, 7
Lometa	Texas		lower 6, 7, 8
Llano	New Mexico		7, 8

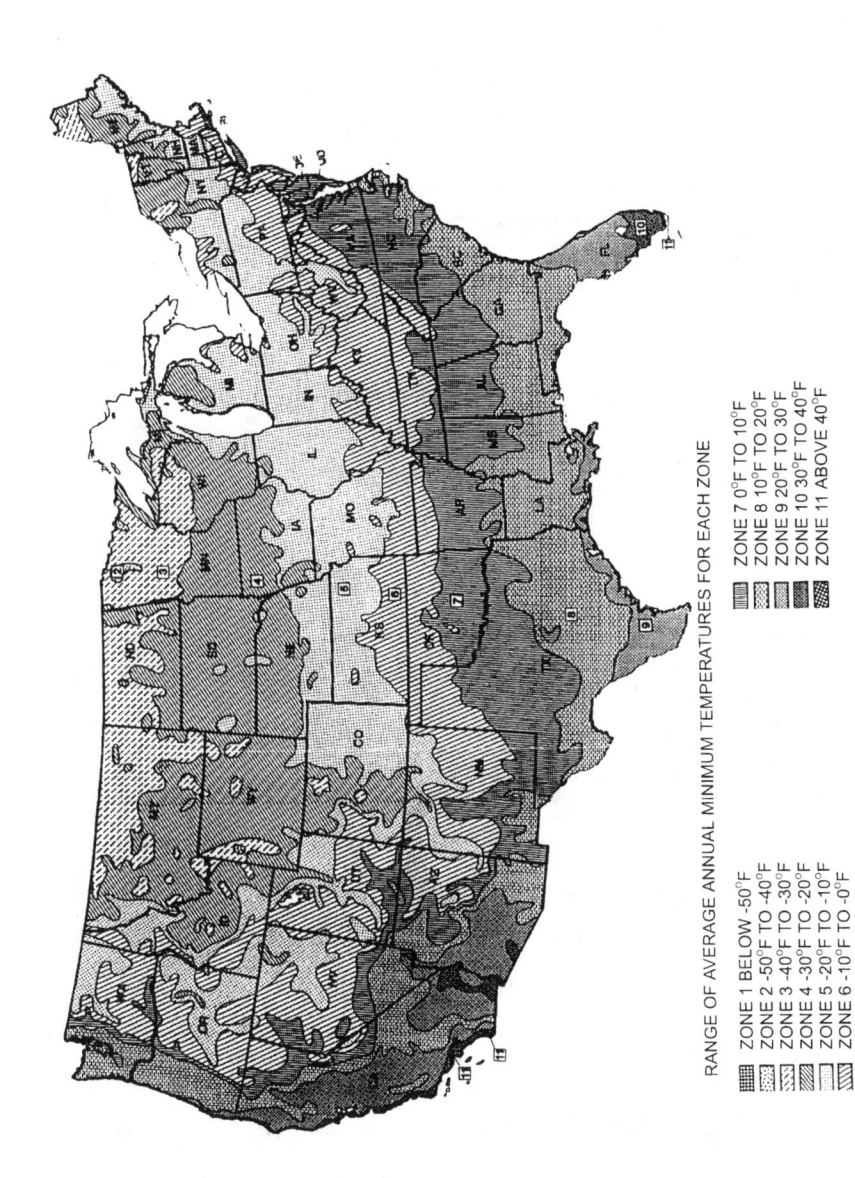

IMPROVING BY USING SELECTION, BREEDING, AND BIOTECHNOLOGY

Fig. 6–5. USDA plant hardiness zone map.

photoperiod response as determined by latitude of origin and length of the growing season are the primary factors affecting cultivar adaptation to specific regions. The released cultivars of these grasses are best adapted and are most productive in areas where annual precipitation exceeds 450 mm.

SUMMARY

Substantial genetic variation exists among and within populations of switchgrass, big bluestem, and indiangrass. Cultivars have been developed for most geographic regions but in comparison to most cultivated species, the breeding work on these species has been limited and substantial genetic improvements are feasible. Improved breeding methods and equipment are available to develop improved cultivars of these species for use in grassland agriculture.

REFERENCES

Allard, R.W. 1964. Principles of plant breeding. John Wiley & Sons, New York.

Anderson, Bruce, J.K. Ward, K.P. Vogel, M.G. Ward, H.J. Gorz, and F.A. Haskins. 1988. Forage quality and performance of yearlings grazing switchgrass strains selected for differing digestibility. J. Anim. Sci. 66:2239–2244.

Alexandrova, K.S., P.D. Denchev, and B.V. Conger. 1996. In vitro development of inflorescences from switchgrass nodal segments. Crop Sci. 36:175–178.

Alderson, J., and W.C. Sharp. 1994. Grass varieties in the United States. USDA-SCS Agric. Handb. 170. U.S. Gov. Print. Office, Washington, DC.

Asay, K.H., and D.A. Johnson. 1987. Breeding for improved seedling establishment in cool-season range grasses. p.173–176. *In* G.W. Frasier and R.A. Evans (ed.) Proc. of Symp. on Seed and Seedbed Ecology of Rangeland Plants, Tucson, AZ. 21–23 Apr. 1987. USDA-ARS Natl. Tech. Information Serv., Springfield, VA.

Asstveit, A.H., and K. Asstveit. 1990. Theory and application of open-pollination and polycross in forage breeding. Theor. Appl. Genet. 79:618–624.

Barnes, P.W. 1986. Variation in the big bluestem (*Andropogon gerardii*)–sand bluestem (*Andropogon hallii*) complex along a local dune/meadow gradient in the Nebraska Sandhills. Am. J. Bot. 73:172–184.

Barnett, F.L., and R.F. Carver. 1967. Meiosis and pollen stainability in switchgrass, *Panicum virgatum* L. Crop Sci. 7:301–304.

Barnett, F.L., Z.E. Rafil, and R.L. Vanderlip. 1971. Seed yield, seed quality, and total establishment capability of indiangrass. Crop Sci. 11:644–647.

Barnett, F.L., and R.L. Vanderlip. 1969. Criteria of field establishment capability in indiangrass, *Sorghastrum nutans* (L.) Nash. Crop Sci. 9:290–293.

Benedict, H.M. 1941. Effect of day length and temperature on the flowering and growth of four species of grasses. J. Agric. Res. 61:661–672.

Boe, A., and P.O. Johnson. 1987. Deriving a large-seeded switchgrass population using air-column separation of parent seed. Crop Sci. 27:147–148.

Boe, A., K. Robbins, and B. McDaniel. 1989. Spikelet characteristics and midge predation of hermaphroditic genotypes of big bluestem. Crop Sci. 29:1433–1435.

Boe, A.A., and J.G. Ross. 1983. Path coefficient analysis of seed yield in big bluestem. J. Range Manage. 36:652–653.

Boe, A.A., J.G. Ross, and R. Wynia. 1983. Pedicellata spikelet fertility in big bluestem from eastern South Dakota. J. Range Manage. 36:131–132.

Bragg, L.H. 1964. Chromosome counts and cytological observations of certain Texas Graimineae under transplanted conditions. Southwest Nat. 9:306–308.

Braverman, S.W. 1986. Disease resistance in cool-season forage and forage range and turf grass II. Bot. Rev. 52:1–112.

Brown, R.R., J. Henry, and J. Crowder. 1983. Improved processing for high quality seedfor big bluestem (*Andropogon gerardii*) and yellow indiangrass (*Sorghastrum nutans*). p.272–274. *In* J.A. Smith and V.W. Hays (ed.) Proc. 14th Int. Grassl. Congr., Lexington, KY. 15–24 June 1981. Westview Press, Boulder, CO.

Brunken, J.N., and J.R. Estes. 1975. Cytological and morphological variation in *Panicum virgatum* L. Southwest Nat. 19:379–385.

Burton, G.W. 1948. The performance of various mixtures of hybrid and parent inbred millet. J. Am. Soc. Agron. 40:908–915.

Burton, G.W. 1974. Recurrent restricted phenotypic selection increases forage yields of Pensacola bahiagrass. Crop Sci. 14:831–835.

Burton, G.W. 1981. Nutrient composition of forage crops. Effects of genetic factors. USDA-ARS Agric. Rev. Manuals. ARM-S-21. USDA, Washington, DC.

Burton, G.W. 1982. Improved recurrent restricted phenotypic selection improves Bahia forage yields. Crop Sci. 22:1058–1061.

Burton, G.W. 1985. Spaced-plant-population-progress test. Crop Sci. 25:63–65.

Burton, G.W. 1986. Developing better forages for the south. J. Anim. Sci. 63:63–65.

Carter, M.R., G.R. Manglitz, M.D. Rethwisch, and K.P. Vogel. 1988. A seed midge pest of big bluestem. J. Range Manage. 41:253–254.

Chen, C.H., and A.A. Boe. 1988. Big bluestem (*Andropogaon gerardii* Vitman), little bluestem [*Schizachyrium scoparius* (Michx.) Nash], and indiangrass [*Sorghastrum nutans* (L.) Nash]. p. 444–457. *In* Y.P.S. Bajaj (ed.) Biotechnology in agriculture and forestry. Vol. 6. Crops II. Springer-Verlag, Berlin.

Church, G.L. 1929. Meiotic phenomena in certain Gramineae: II. Paniceae and Andropogoneae. Bot. Gaz. 88:63–64.

Clements, R.J., and B.D.H. Latter. 1973. Responses to selection for seed weight and seedling vigor in phalaris. Aust. J. Agric. Res. 25:33–34.

Cornelius, D.R. 1950. Seed production of native grasses. Ecol. Mono. 20:1–27.

Cornelius, D.R., and C.O. Johnston. 1941. Differences in plant type and reaction to rust among several collections of *Panicum virgatum* L. J. Am. Soc. Agron. 33:115–124.

Denchev, P.D., and B.V. Conger. 1994. Plant regeneration from callus culture of switchgrass. Crop Sci. 34:1623–1627.

Dewald, G.W., and S.M. Jahal. 1977. Meiotic behavior and fertility interrelationships in *Andropogon scoparius* and *A. gerardii*. Cytologia. 39:214–223.

Eberhardt, S.A., and L.C. Newell. 1959. Variation in domestic collections of switchgrass, *Panicum virgatum* L. Agron. J. 51:613–616.

Emal, J.G., and E.C. Conard. 1973. Seed dormancy and germination in indiangrass as affected by light, chilling, and certain chemical treatments. Agron. J. 81:47–52.

Falconer, D.S. 1981. Introduction to quantitative genetics 2nd ed. Longman, New York.

Gabrielsen, B.C., K.P. Vogel, B.E. Anderson, and J.K. Ward. 1990. Alkali-labile lignin phenolics and forage quality in three switchgrass strains selected for differing digestibility. Crop Sci. 30:1313–1320.

Glewen, K.L., and K.P. Vogel. 1984. Partitioning the genetic variability for seedling growth in sand bluestem into its seed size and seedling vigor components. Crop Sci. 24:137–141.

Godshalk, E.B., J.C. Burns, and D.H. Timothy. 1986. Selection for in vitro dry matter digestibility disappearance in switchgrass regrowth. Crop Sci. 26:943–947.

Godshalk, E.B., and D.H. Timothy. 1988a. Factor and principal component analyses as alternatives to index selection. Theor. Appl. Genet. 76:352–360.

Godshalk, E.B., and D.H. Timothy. 1988b. Effectiveness of index selection for switchgrass forage yield and quality. Crop Sci. 28:825–830.

Godshalk, E.B., W.F. McClure, J.C. Burns, D.H. Timothy, and D.S. Fisher. 1988a. Heritability of cell wall carbohydrates in switchgrass. Crop Sci. 28:736–742.

Godshalk, E.B., D.H. Timothy, and J.C. Burns. 1988b. Application of multistage selection indices to crop improvement. Crop Sci. 28: 23–26.

Gorz, H.J., F.A. Haskins, R. Dam, and K.P. Vogel. 1979. Dhurrin in *Sorghastrum nutans*. Phytochemistry 18:20–24.

Gould, F.W. 1968. Grass systematics. McGraw-Hill Book Co., New York.

Gould, F.W. 1975. The grasses of Texas. Texas A&M Univ. Press, College Station.

Gunter, L.E., G.A. Tuskan, and S.D. Wullschleger. 1996. Diversity among populations of switchgrass based on RAPD markers. Crop Sci. 36:1017–1022.

Halpin, C., M.E. Knight, B.A. Foxon, M.M. Campbell, A.M. Boudet, J.J. Boon, B. Chabbert, M-T. Tollier, and W. Schuch. 1994. Manipulation of lignin quality by down regulation of cinnamyl alcohol dehydrogenase. Plant J. 6:339–350.

Hallauer, A.R., and J.B. Miranda, OF. 1981. Quantitative genetics in maize breeding. Iowa State Univ. Press, Ames.

Hanson, A.A., and H.L. Carnahan. 1956. Breeding perennial forage grasses. USDA-ARS Agric. Tech. Bul. 1145. USDA, Washington, DC.

Hitchcock, A.S. 1971. Manual of grasses of the United States. 2nd ed. Dover Publ., New York.

Hopkins, A.A., K.P. Vogel, and K.J. Moore. 1993. Predicted and realized gains from selection for *in vitro* dry matter digestibility and forage yield in switchgrass. Crop Sci. 32:253–258.

Hopkins, A.A., K.P. Vogel, K.J. Moore, K.D. Johnson, and I.T. Carlson. 1995a. Genetic effects and genotype by environment interactions for traits of elite switchgrass populations. Crop Sci. 35:125–132.

Hopkins, A.A., K.P. Vogel, K.J. Moore, K.D. Johnson, and I.T. Carlson. 1995b. Genetic variability and genotype x environment interactions among switchgrass accessions from the Midwestern USA. Crop Sci. 35:565–571.

Hopkins, A.A., C.M. Taliaferro, C.D. Murphy, and D. Christian. 1996. Chromosome number and nuclear DNA content of several switchgrass populations. Crop Sci. 36:1192–1195.

Hultquist, S.J., K.P. Vogel, D.J. Lee, K. Arumuganathan, and S. Kaeppler. 1996. Chloroplast DNA and nuclear DNA content variations among cultivars of switchgrass, *Panicum virgatum* L. Crop Sci. 36:1049–1052.

Hultquist, S.J., K.P. Vogel, D.E. Lee, K. Arumuganathan, and S. Kaeppler. 1997. Nuclear DNA content and chloroplast DNA polymorphisms among accessions of *Panicum virgatum* L. from remnant Midwestern prairies. Crop Sci. 37:595–598.

Johnson, P., and A.A. Boe. 1982. Seed size variation in three switchgrass (*Panicum virgatum* L.) varieties. Proc. S. Dak. Acad. Sci. 61:159.

Jones, M.D., and L.C. Newell. 1946. Pollination cycles and pollen dispersal in relation to grass improvement. Nebraska Agric. Exp. Stn. Res. Bull. 148. Nebraska Agric. Exp. Stn., Lincoln.

Jones, M.D., and L.C. Newell. 1948. Size, variability, and identification of grass pollen. J. Am. Soc. Agron. 40:136–143.

Kassel, P.C., R.E. Mullen, and T.B. Bailey. 1985. Seed yield response of three switchgrass cultivars for different management practices. Agron. J. 77:214–218.

Keeler, K.H. 1990. Distribution of polyploid variation in big bluestem (*Andropogon gerardii*) across the tallgrass prairie region. Genome 33:95–100.

Keeler, K.H. 1992. Local polyploid variation in the native prairie grass *Andropogon gerardi*. Am. J. Bot. 79:1229–1232.

Kneebone, W.R. 1956. Breeding for seedling vigor in sand bluestem (*Andropogon hallii* Hack.) as estimated from parental clones and their open-pollinated progenies. Agron. J. 50:459–461.

Kube, J.G., K.P. Vogel, and L.E. Moser. 1989. Genetic variability for seedling atrazine tolerance in indiangrass (*Sorghastrum nutan*). Crop Sci. 29:18–23.

Law, A.G., and K.L. Anderson. 1940. The effect of selection and inbreeding on the growth of big bluestem (*Andropogon furcatus* Muhl.). J. Am. Soc. Agron. 32:931–943.

Lu, Ku, S.M. Kaeppler, K.P. Vogel, and K. Arumuganathan. 1998. Nuclear DNA content and chromosome numbers on switchgrass. Great Plains Res. 8:269–280.

Marten, G.C. 1989. Breeding forage grasses to maximize animal performance. p. 71–104. *In* D.A. Sleper et al. (ed.) Contributions from breeding forage and turf grasses. CSSA Spec. Publ. 15. CSSA, Madison, WI.

McKell, C.M. 1972. Seedling vigor and seedling establishment. p. 76–89. *In* V.B. Younger and C.M. McKell (ed.). The biology and utilization of grasses. Academic Press, New York.

McKendrick, J.D., C.E. Owensby, and R.M. Hyde. 1975. Big bluestem and indiangrass vegetative reproduction and annual reserve carbohydrate and nitrogen cycles. Agric. Ecosyst. 2:75–93.

McMillian, C. 1959. The role of ecotypic variation in the distribution of the central grassland of North Am. Ecol. Mono. 29:285–308.

McMillan, C. 1965. Ecotypic differences within four North American prairie grasses: II. Behavioral variation within transplanted community fractions. Am. J. Bot. 52:55–65.

Moser, L.E., and K.P. Vogel. 1995. Switchgrass, big bluestem, and indiangrass. p. 409–420. *In* R.F Barnes et al. (ed.). Forages. 5th ed. Vol. I. An introduction to grassland agriculture. Iowa State Univ. Press, Ames.

Nelson, C.J., D.A. Sleper, and J.H. Coutts. 1985. Field performance of tall fescue selected for leaf-area expansion. p. 320–322. *In* Proc. of the XV Int. Grassland Congr., Kyoto, Japan. 24–31 Aug. 1985. Japanese Soc. of Grassland Sci. and the Science Council of Japan, Tishi-nasuno, Tachigi-ken, Japan.

Newell, L.C. 1968. Effects of strain source and management practice on forage yields of two warm-season prairie grasses. Crop Sci. 8:205–210.
Newell, L.C., and S.A. Eberhardt. 1961. Clone and progeny evaluation in the improvement of switchgrass. Crop Sci. 1:370–373.
Newell, L.C., and L.V. Peters. 1961. Performance of hybrids between divergent types of big bluestem and sand bluestem in relation to improvement. Crop Sci. 1:370–373.
Nielsen, E.L. 1944. Analysis of variation in *Panicum virgatum*. J. Agric. Res. 69:327–358.
Norrmann, G.L., C.L. Quarin, and K.H. Keeler. 1997. Evolutionary implications of meiotic chromosome behavior, reproductive biology, and hybridization in 6x and 9x cytotypes of *Andropogon gerardii* (Poaceae). Am. J. Bot. 84:201–207.
Pedersen, J.F., R.D. Lee, D.J. Lee, and K.P. Vogel. 1993. Comparison of sorghum and indiangrass chloroplast genomes using RFLPs. p. 196–2000. *In* Proc. American Forage and Grassland Council, Des Moines, IA. 29–31 Mar. 1993. Am. Forage and Grassland Council, Georgetown, TX.
Pedreira, C.G.S., and R.H. Brown 1996. Physiology, morphology, and growth of individual plants of selected and unselected bahiagrass populations. Crop Sci. 36(1):138–142.
Peters, L.V., and L.C. Newell. 1961. Hybridization between divergent types of big bluestem, *Andropogon gerardii* Vitman, and sand bluestem, *Andropogon hallii* Hack. Crop Sci. 1:359–363.
Porter, C.L. 1966. An analysis of variation between upland and lowland switchgrasses, *P. virgatum*, in central Oklahoma. Ecology 47:980–992.
Pouli, J.R., R.L. Reid, and D.P. Belesky. 1992. Photosensitization in lambs grazing switchgrass. Agron. J. 84:1077–1080.
Riley, R.D. 1981. Heritability of mature plant traits in sand bluestem. Ph.D. diss. Univ. of Nebraska, Lincoln.
Riley, R.D., and K.P. Vogel. 1982. Chromosome numbers of released cultivars of switchgrass, indiangrass, big bluestem, and sand bluestem. Crop Sci. 22:1081–1083.
Ross, J.G., R.T. Thaden, and W.L. Tucker. 1975. Selection criteria for yield and quality in big bluestem (*Andropogon gerardi* Vitman). Crop Sci. 15:303–306.
Schumacher, C.M. 1962. Grass production in Nebraska and South Dakota. USDA-SCS Technical Guide-Section IV-G. USDA, Washington, DC.
Sebolai, B., and K.P. Vogel. 1989. Evaluation of three cycles of mass selection for seedling weight in switchgrass, indiangrass, and big bluestem. p. 99. *In* Agronomy abstracts. ASA, Madison, WI.
Snetselaar, K.M., and L.H. Tiffany. 1990. Light and electron microscopy of sorus development in *Sorosporium provinciale*, a smut of big bluestem. Mycologia 82:480–492.
Stubbendieck, J., S.L. Hatch, and C.H. Butterfield. 1991. North American range plants. Univ. of Nebraska Press, Lincoln.
Talbert, L.E., D.H. Timothy, J.C. Burns, J.O. Rawlings, and R.H. Moll. 1983. Estimates of genetic parameters in switchgrass. Crop Sci. 23:725–728.
Trupp, C.R. and I.T. Carlson. 1971. Improvement of seedling vigor of smooth bromegrass (*Bromus inermis* Leyss.) by recurrent selection for high seed weight. Crop Sci. 11:225–228.
Vogel, K.P. 1996. Energy production from forage or American agriculture: Back to the future. J. Soil Water Conserv. 51:137–139.
Vogel, K.P., and B.C. Gabrielsen. 1986. Breeding to improve native warm-season grasses. p. 27–34. *In* Warm-season grasses. Balancing forage programs in the northeast and southern Corn Belt. Soil Conserv. Soc. of Am., Ankeny, IA.
Vogel, K.P., R. Britton, H.J. Gorz, and F.A. Haskins. 1984. *In vitro* and *in vivo* analyses of hays of switchgrass strains selected for high and low IVDMD. Crop Sci. 24:977–980.
Vogel, K.P., H.J. Gorz, and F.A. Haskins. 1981a. Heritability estimates of forage yield, *in vitro* dry matter digestibility, crude protein, and heading date in indiangrass. Crop Sci. 21:35–38.
Vogel, K.P., H.J. Gorz, and F.A. Haskins. 1981b. Divergent selection for *in vitro* dry matter digestibility in switchgrass. Crop Sci. 21:39–41.
Vogel, K.P., H.J. Gorz, and F.A. Haskins. 1981c. Heritability estimates for height, color, erectness, leafiness, and vigor in indiangrass. Crop Sci. 21:734–736.
Vogel, K.P., H.J. Gorz, and F.A. Haskins. 1989. Breeding grasses for the future. p. 105–122. *In* D.A. Sleper et al. (ed.) Contributions from breeding forage and turf grasses. CSSA Spec. Publ. 15. CSSA. Madison, WI.
Vogel, K.P., F.A. Haskins, and H.J. Gorz. 1980. Parent-progeny regression in indiangrass: Inflation of heritability estimates by environmental covariances. Crop Sci. 20:580–582.
Vogel, K.P., F.A. Haskins, and H.J. Gorz. 1987. Potential for hydrocyanic acid poisoning of livestock by indiangrass (*Sorghastrum nutans*). J. Range Manage. 40:506–509.

Vogel, K.P., A.A. Hopkins, K.J. Moore, K.D. Johnson, and I.T. Carlson. 1996. Registration of 'Shawnee' switchgrass. Crop Sci. 36:1713.

Vogel, K.P. and G.R. Manglitz. 1989. Effect of the big bluestem seed midge on the sexual reproduction of big bluestem: a review. p. 267–291. *In* T.B. Bragg and J. Stubbendieck (ed.) Proc. of the 11th North American Prairie Conf., Prairie Pioneers, Ecology, History, and Culture. Lincoln, NE. 7–11 Aug. 1988. Univ. of Nebraska, Lincoln.

Vogel, K.P., and G.R. Manglitz. 1990. Evaluation of Furadan and orthene against a bluestem seed midge, 1986 and 1987. Insect. Acaracide Tests 15:175–176.

Vogel, K.P., K.J. Moore, and A.A. Hopkins. 1993. Breeding switchgrass for improved animal performance. p. 1734–1735. *In* Proc. XVII Int. Grassland Congr., Palmerston North, Hamilton, Lincoln and Rockhampton, Australia, 8–21 Feb. 1993. New Zealand Grassland Society, Tropical Grasslands Soc. of Australia, New Zealand Soc. of Animal Production, Australian Soc. of Animal Production, and New Zealand Inst. of Agric. Sci.

Vogel, K.P., K.J. Moore, K.D. Johnson, and I.T. Carlson. 1994. Variation among indiangrass from Midwestern remnant prairies for agronomic traits. p. 220. *In* Agronomy abstracts. ASA, Madison, WI.

Vogel, K.P., and J.F. Pedersen. 1993. Breeding systems for cross-pollinated perennial grasses. Plant Breed. Rev. 11:251–274.

Vogel, K.P. and D.A. Sleper. 1994. Alteration of plants via genetics and plant breeding. p. 891–921. *In* G.C. Fahey, Jr. (ed.). Forage quality, evaluation, and utilization. ASA, CSSA, SSSA, Madison, WI.

Voight, P.W., C.R. Tischler, and B.A. Young. 1987. Selection for improved establishment in warm-season grasses. p. 177–187. *In* G.W. Frasier and R.A. Evans (ed.) Proc. Symp., Seed and Seedbed Ecology of Rangeland Plants, Tucson, AZ. 21–23 Apr. USDA-ARS, Washington, DC.

Vybiral, A.J., K.P. Vogel, K.J. Moore, K.D. Johnson, and I.T. Carlson. 1993. Variation among big bluestem from Midwestern remnant prairies for agronomic traits. p. 196. *In* Agronomy abstracts. ASA, Madison, WI.

Weaver, J.E. 1968. Prairie plants and their environment. Univ. of Nebraska Press, Lincoln.

Weeks, J.T., O.D. Anderson, and A.E. Blechl. 1993. Rapid production of multiple independent lines of fertile transgenic wheat (*Triticum aestivum*). Plant Physiol. 102:1077–1084.

Weimer, M.R., B.A. Swisher, and K.P. Vogel. 1988. Metabolism as a basis for inter- and intra-specific atrazine tolerance in warm-season grasses. Weed Sci. 36:436–440.

Whalley, R.D.B., C.M. McKell, and L.R. Green. 1966. Seedling vigor and the non-photosynthetic stage of seedling growth in grasses. Crop Sci. 6:147–150.

Wheeler, W.A., and D.D. Hill. 1957. Grassland seeds. D. Van Norstrand Co., Princeton, NJ.

Wright, L.N. 1977. Recurrent selection for shifting gene frequency of seed weight in *Panicum antidotale* Retz. Crop Sci. 17:176–178.

Zeiders, K.E. 1982. Leaf spots of big bluestem, little bluestem, and indiangrass caused by *Ascochyta brachypodii*. Plant Dis. 66:502–505.

Zeiders, K.E. 1984. Helminthosporium spot blotch of switchgrass (*Panicum virgatum*) in Pennsylvania. Plant Dis. 68:120–122.

7 An Overview of Seed Dormancy in Native Warm-Season Grasses

Allen D. Knapp

Iowa State University
Ames, Iowa

Stands of native, warm-season grasses can be difficult to establish from seed. Problems in stand establishment are often attributed to seed dormancy while, in other cases, causal factors are believed to be related to slow seedling growth, slow development of adventitious roots or low seed quality. Additionally, seed laboratories often have difficulty in assessing the planting value of these species and problems arise in both the physical and physiological assessments of seed quality. Some of these problems derive from the structures encompassing the caryopsis that often comprise part of the dispersal unit, from the pericarp and testa, and still others from the physiological status of the embryo itself as potentially modified by the encompassing structures including the pericarp and testa. Taken together, these issues impact all potential users of native warm-season grass species from the producer to the consumer. In order to maximize the successful use of a species, consumers must be confident that following applicable guidelines will result in a reasonable probability of successful establishment.

This is an overview of seed dormancy in certain native, warm-season grasses. It will not consider other aspects of seed quality that may, in some ways, mimic outcomes often attributed to seed dormancy. Certainly, seed dormancy is a fundamental limitation to the establishment of many of these species, however, it should not be confused with stand establishment problems associated with low viability or vigor.

For the purposes of this overview, no differentiation will be made between germination per se and seedling growth. While it may be valid to separate the establishment process into germination (those processes leading up to the initiation of seedling growth) from seedling development, there is certainly the potential for the legacy of germination physiology to affect subsequent seedling development. Further, the germination tests used to provide labeling information and a majority of literature related to seed germination include a normal seedling concept in the definition of germination. In reference to discussions of laboratory data, therefore, the term germination will be used even though subsequent seedling development also may have been evaluated. The term seed will be applied to the dispersal unit of a species even though this unit may include florets, spikelets and other structures.

Copyright © 2000. Crop Science Society of America and American Society of Agronomy, 677 S. Segoe Rd., Madison, WI 53711, USA. *Native Warm-Season Grasses: Research Trends and Issues.* CSSA Special Publication no. 30.

SEED DORMANCY CONCEPTS

Definitions

Few other aspects of seed and seed performance have captured as much attention in the scientific arena as has seed dormancy. The agronomic, horticultural, ecological and general biological implications of seed dormancy are far reaching. An organism's ability to cease development while maintaining viability, often for extended periods of time, is engaging and important to many biologists. There are, therefore, several potential approaches to the topic of seed dormancy ranging from the basic molecular biology of the condition to a practical assessment of how to manage dormancy in species that may produce dormant seed. The interest in seed dormancy by diverse research disciplines has resulted in several good reviews of the topic. No attempt will be made here to discuss dormancy in the detail available in these reviews. Instead, the reader will be referred to these works in the course of this chapter.

A substantial diversity exists in the terminology used to describe and define seed dormancy also exists. It appears that this diversity is the result of an inability to agree upon either descriptive or mechanistic approaches. Thus, no definition of dormancy has been generally agreed upon and the literature is replete with definitions and vague classifications attempting to describe a particular dormancy behavior. Usually, these classifications or definitions confound a description of the germination state and the believed origin of the dormancy or the environmental conditions in which the dormancy is exhibited. The definition of dormancy is a formidable task but since the basic physiological mechanisms are usually, as yet, poorly understood, a predominately descriptive definition probably is most reasonable.

Seed dormancy has been defined (Villiers, 1972) as "the state of arrested development whereby the organ or organism, by virtue of its structure or chemical composition, may possess one or more mechanisms preventing its own germination." This definition is elegant in its brevity yet remarkably complete in its description of dormancy. Villiers' (1972) definition indicates that dormancy is a result of an organ or organism's structure or chemical composition. This implies that dormancy is derived from the seed itself, that it "may possess one or more mechanisms preventing its own germination." According to this definition, it is the structural or physiological status of the seed that results in arrested development. While it may be more valid to use ". . . structural and chemical composition . . .", acceptance of the concept that dormancy arises from the seed (dispersal unit) itself begins to clarify what terminology should be used to discuss seed dormancy. The interactions between the seed and the particular environment that the seed is exposed to, however, are important.

Villiers' (1972) definition of dormancy does not provide for a lack of germination due to unfavorable environmental conditions. Villiers defines the latter situation as quiescence or "a state of arrested development maintained solely by unfavorable environmental conditions such as inadequate water supply." Functionally then, a viable, dry seed may be dormant or quiescent. Its germination capabilities cannot be determined without first providing conditions favorable for germination. If the seed germinates following the provision of adequate moisture,

OVERVIEW OF SEED DORMANCY

an appropriate temperature, appropriate oxygen level, and (perhaps) light, it was quiescent. If the viable seed does not germinate even after provision of environmental requirements for germination, it is dormant.

Another important attribute of seed performance in relation to dormancy is the potential of nondormant seed of some species to become dormant. This attribute has been classified as secondary dormancy (Villiers, 1972). Specifically, Villiers (1972) has defined secondary dormancy as "those cases where dormancy becomes imposed on seeds by imbibition under conditions unfavorable to germination so that seeds become unable to germinate at temperatures normally favorable for germination of that species until some special releaser stimulus is applied." Villiers' (1972) definition of secondary dormancy seems to require the imposition or development of specific germination block(s) in response to these *unfavorable* conditions as a prerequisite for the use of the term dormancy. Certainly, this is consistent with the definition of quiescence. This distinction also is important practically since a seed rendered quiescent by unfavorable environments might be able to germinate and establish promptly once the unfavorable environmental conditions are removed or changed. Alternatively, seed that have entered a secondary dormancy as a result of exposure to unfavorable environmental conditions would not germinate until some *special releaser stimulus is applied*.

Villiers' definitions provide a simple and useful framework for evaluating and managing seed dormancy. There are, however, a number of different approaches to defining seed dormancy. One such definition (Chancellor, 1984), states "Seeds which are alive but do not germinate are dormant. This dormancy can be caused either by characteristics of the seeds themselves or by the environment in which they occur." Whether Chancellor's definition requires a metabolic shift of sufficient consequence to be manifest even under ideal conditions is unclear. Chancellor's definition seems further to indicate that dormancy is not necessarily derived from the seed. The definitions of Villiers (1972) and Chancellor (1984) illustrate a fundamental question. Specifically, is the state of arrested development brought about by the lack of appropriate environmental variables the same as that which renders a viable seed incapable of germinating when the appropriate environmental variables are provided? An obvious answer to that question is no. The differential behavior of seeds in environments normally considered conducive to germination indicates that, at least quantitatively, the arrested states are different. This issue should not be oversimplified. Research involving a wide range of species indicates that seed dormancy is not an all or none phenomenon. Rather, this literature suggests that dormancy develops and dissipates in a describable fashion. For further discussion of definitions of dormancy see discussions in Villiers (1972); Khan (1977); and Bewley and Black (1982).

Development and Dissipation of Dormancy

Vegis (1964) described the development and dissipation of arrested growth. His (Vegis, 1964) description of an organ's dormancy process involves three states: predormancy, true dormancy, and post dormancy. Vegis discusses these states in relation to growth activities within given environmental ranges. Accordingly, if a seed is not in one of the three dormancy or resting states, the environmental range across which growth and development can proceed is the widest possible. Vegis proposes that the *width* of this environmental range is a function of the species under consideration, thereby considering species adaptation in his description (Fig. 7–1). As the seed progresses through predormancy, the range of environmental conditions under which germination occurs narrows. During this process, Vegis indicates that the seed is in a condition of relative dormancy because germination can be obtained but only within certain environmental parameters. The next phase or state is true dormancy and, according to Vegis (1964), the seed cannot germinate during this period regardless of the environment. He indicates, however, that not all species go through the true dormancy phase instead, some pass directly from predormancy to post dormancy. This means that some species retain the ability to germinate throughout the dormancy cycle although precise environmental conditions might be required for this to occur. The final stage of dormancy is called post dormancy. During this period the range of environmental conditions within which germination can occur widens. When germination will occur within the entire range of environments to which the species is adapted, the seed is no longer dormant. Lang (1965) also emphasizes the importance of environmental conditions in determining the relative germinability of seed under specific environmental conditions. He believes it is a common error to assume that seeds able to germinate under a given set of environmental parameters are not dormant. These works describe a dynamic dormancy system that is logically a function of the species and the environment.

Amen (1968) also published a model of seed dormancy. Amen proposed that dormancy phenomena consist of four mechanisms, which he termed: induction, maintenance, trigger, and germination. The induction mechanism supports the development or onset of dormancy that is maintained until some triggering agent ("that elicits germination, but whose continued presence is not essential") promotes a shift

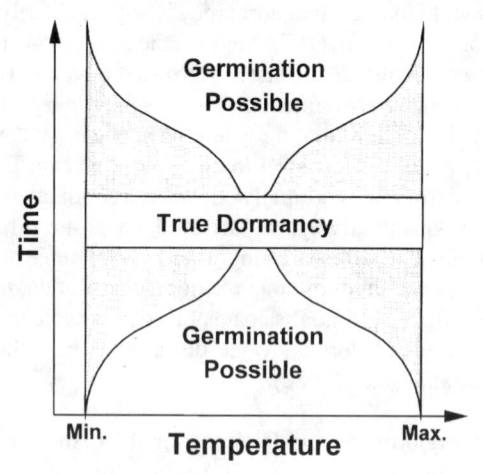

Fig. 7–1. The development and dissipation of seed dormancy as related to time and the range of environmental conditions (for simplicity temperature is used) within which germination can occur, adapted from Vegis (1964). The shaded areas represent the range of environmental conditions, normally conducive to germination of a given species, within which germination of dormant seed will not occur. The shaded area closest to the origin would represent Vegis's pre-dormancy phase and the shaded area, distal from the origin and beyond the true dormancy phase represents Vegis's post-dormancy phase.

in inhibitor–promoter levels in favor of the promoter. This shift favors the dissolution of dormancy and progress toward germination. Amen's model highlights consideration of a process by which seeds become dormant and, subsequently, become nondormant.

As previously mentioned, Amen's (1968) model is based on growth promoter and inhibitor balances. Currently, there is substantial debate regarding the role of naturally occurring plant growth regulators (PGRs) such as abscisic acid (ABA) and gibberellins (GAs) in the control of seed dormancy. Earlier work was limited to exogenous application of PGRs. While this early work substantiated that PGRs could indeed influence the germination of many species, questions regarding the comparability of the actions of exogenously applied PGRs to the action of endogenous compounds arose. With the application of gas chromatography and immunoassay procedures, more precise estimates of the levels of these naturally occurring PGRs could be made. Many of these studies, however, are correlative in nature and the role of PGRs in the control of seed and fruit development, precocious germination, dormancy, and others, remains unclear. Recent research using hormone biosynthesis and response mutants are, however, now providing valuable information regarding the possible modes of action of naturally occurring PGRs (Karssen, 1995; Kermode, 1995).

Villiers (1972) and Lang (1965) also introduce temporal and environmental variables that are important in understanding the germinability of seed of a given species at a given point in time. Simpson (1990) incorporates temporal and environmental variables in his definition of dormancy: "Temporary failure of a viable seed to germinate, after a specified length of time, in a particular set of environmental conditions that later evoke germination when the restrictive state has been terminated by either natural or artificial means." This dormancy definition along with Simpson's (1990) systems approach to analyzing dormancy provide a good basis for understanding that dormancy changes over time as a result of the interactions between dormancy controlling variables and the environment.

Dormancy Mechanisms

Any discussion of dormancy mechanisms must be preceded with a disclaimer indicating that the specific causes or mechanisms of dormancy are not fully understood at this time. This is not to say that little is known about the physiological basis of seed dormancy. Several books referencing or reviewing dormancy have been published (Villiers, 1972; Heydecker, 1973; Khan, 1977, 1982). For a detailed discussion of the physiology of seed dormancy, the reader is referred to these excellent works that detail many issues regarding the basic physiological mechanisms that may control dormancy. In this section, dormancy mechanisms will be considered at the organ or organism level. It is hoped this will be of value to those beginning to evaluate seed dormancy in species that have had little work done on them.

In presenting dormancy mechanisms at the organ or organism level, the author agrees with Villiers (1972) in citing that an older, relatively simple description, has served well for many years (Crocker, 1916). Crocker separates dormancy mechanisms into seven categories. The following are direct quotes of the mechanisms described by Crocker:

1. Rudimentary embryos that must mature before germination can begin;
2. Complete inhibition of water absorption;
3. Mechanical resistance to the expansion of the embryo and seed contents by enclosing structures;
4. Encasing structures interfering with oxygen absorption by the embryo and perhaps carbon dioxide elimination from it;
5. A state of dormancy in the embryo itself or some organ of it, in consequence of which it is unable to grow when naked and supplied with all ordinary germinative conditions;
6. Combinations of two or more of these;
7. Assumption of secondary dormancy.

These categories cover a broad range of dormancy types and provide a reasonable starting point for investigating the manifestations of dormancy in a given species.

A quick review of Crocker's (1916) list of dormancy mechanisms provides several important concepts. First, dormancy is primarily expressed in two ways, via the structures encompassing the embryo and via mechanisms within the embryo itself. Another important concept is presented in mechanism six, which indicates that some seed species may possess more than one dormancy mechanism. Not quite so apparent from the list is the potential interaction between the encompassing structures and the physiology of the embryo itself. Also, while the potential occurrence is referenced in the article, there is no direct representation in the preceding list of the possible occurrence of germination inhibiting substances in the structures encompassing the seed. Interactions between the encompassing structures and the physiological status of the seed and combinations of dormancy types provide substantial challenges to those trying to understand the biology of seed dormancy as well as those trying to establish treatments to consistently produce rapid germination in dormant seed lots.

SEED DORMANCY IN CERTAIN NATIVE GRASSES

Determining the germination behavior of a species within its entire range of adaptation is a formidable task. Thus, few species have been studied in the detail necessary to gain a complete knowledge of the range of environmental, structural, and physiological variables controlling germination. Fortunately, this is not often required in order to manage dormancy in general agronomic practice.

There appear to be substantial similarities among certain native warm-season grasses regarding germination and dormancy parameters. A first indication of these similarities is obtained by reviewing the standard germination requirements of these species as defined by the Association of Official Seed Analysts (AOSA, 1993) (Table 7–1). It also appears that dormant seed of several of the native warm-season grass species germinate best in response to some combination of scarification and chemical or thermal treatment.

Some of the earliest research on seed germination of native warm-season grasses addressed the effects of storage on the germination of these species. Cuokos

Table 7–1. Germination conditions for given warm-season grass species adapted from the Association of Official Seed Analysts.

Species	Substrata†	Temp.‡	First count	Final count	Specific requirements	Fresh and dormant
Schizachyrium scoparium	P, TS	20–30	7	14	Light, KNO$_3$	Prechill at 5°C for 2 wks
Sorghastrum nutans	P, TS	20–30	7	14	Light, KNO$_3$	Prechill at 5°C for 2 wks
Bouteloua gracillus	P, TB	20–30	7	14	Light	KNO$_3$
Bouteloua curtipendula	P	15–30	7	14	Light, KNO$_3$	
Buchloe dactyloides						
burs	P, TB, TS	20–35	7	14	Light, KNO$_3$	Prechill at 5°C for 2 wks
caryopses	P	20–35	5	14	Light, KNO$_3$ Optional	
Panicum virgatum	P, TS	15–30	7	14	Light, KNO$_3$	Prechill at 5°C for 2 wks
Andropogon gerardii var. *gerardii*	P, TS	20–30	7	14	Light, KNO$_3$	Prechill at 5°C for 2 wks

† P, covered petri dishes; TS, top of sand; TB, top of blotters.
‡ Temperature in °Celsius.

(1944), citing the need for investigations into the reasons for seeding failures experienced in attempts to establish native grass stands, published the results of a seed storage study. Seed of big bluestem (*Andropogon gerardii* Vit.), little bluestem (*Andropogon scoparius* Michx.), indiangrass [*Sorghastrum nutans* (L.) Nash.], and sideoats grama (*Bouteloua curtipendula* Michx.) were placed into different general storage environments including cold storage, room temperature, and a barn loft (Coukos, 1944). The seed dormancy of the species investigated was classified as prolonged or brief based on whether the dormancy persisted into the first spring after harvest or for 2 to 3 mo, respectively. Germination tests consisted of duplicate 100-seed samples in a sterilized peat and sand mixture observed in the greenhouse for 4 wk. Based on these criteria, Coukos indicated that seed of big bluestem, indiangrass, little bluestem, and some ecotypes of sideoats grama possessed prolonged dormancy but that the duration of dormancy and viability depended on conditions in which the seed were stored and perhaps on the *strain* or location of production. The beneficial effects of storage on the germination of these seeds was confirmed in the field where it was found that freshly harvested seed produced thin stands but 2-yr-old seed produced *comparatively good stands*.

Seed of Syn-1 SD 32 switchgrass seed were harvested in 1978, 1979, 1980, and 1981 and placed into cold storage (Boe & Wynia, 1982). Germination increased with seed age. Results of germination tests conducted at constant temperatures of 15 and 25°C and at an alternating temperature of 15/30°C (16/8 hr) were reported. Three-year-old seed germinated 48, 69, and 69% at 15, 25, and 15/30°C, respectively. Comparable germination values for freshly harvested seed were, 1, 2, and 23% at 15, 25, and 15 to 30°C, respectively. Cold storage of seed from the 1981 harvest for an additional 2 mo significantly increased seed germination at the alternating temperature. In a storage study involving caryopses of buffalograss [*Buchloe dactyloides* (Nutt.) Engelm.], seed size and year by seed size variances were not significant factors relative to seed longevity (Kneebone, 1960). Germination of the caryopses of buffalograss, however, was affected by age and increased from 1952 to 1956 then declined rapidly to low levels by 1959. Shaidee, et al. (1969) found that the age of seed substantially influenced both laboratory germination and field emergence. The age of seed that produced the best field emergence varied among the varieties of lots tested. In this study (Shaidee et al., 1969), 1-yr-old seed of sand bluestem (*Andropogon hallii* Hack.), blue grama [*Bouteloua gracilis* (HBK.) Lag.], and A-6606 switchgrass (*Panicum virgatum* L.) and 2-yr-year old seed of sideoats grama and yellow indiangrass (*Sorghastrum nutans* L. Nash.) provided the best field emergence. Interestingly, a 7-yr-old sample of 'Grenville' switchgrass produced the highest field emergence of the three lots of that variety tested. Shaidee et al. (1969) also found substantial variation between the laboratory germination and field emergence in some of the lots and species tested. Specifically, while some of the younger seed lots germinated better than older lots in the laboratory, older seed lots sometimes did better in the field. They hypothesized that while dormancy was broken, some factors in the field might reintroduce dormancy resulting in lower field stands and that the younger seed lots might be more susceptible to this secondary dormancy.

These studies provide practical evidence for the variation in germination response over time. As well, they point out a potential influence of cultivar, year of production and storage conditions on the dormancy levels of grass seed. The effects of year of production on germination are not well understood in native warm-season grasses but could range from the influence of the maturation environment on the expression of the inherent genetic potential for dormancy to factors influencing seed vigor.

The influence of seed size on germination and vigor have been studied in several warm-season grass species including buffalograss, indiangrass, sand bluestem, sideoats grama, and switchgrass (Kneebone & Cremer, 1955). Kneebone & Cremer (1955) conducted studies during a period of 3 yr and indicated that large seed generally produced more vigorous seedlings based on emergence rate, a vigor score, stand, height, and forage production. Seed size did not have a major affect on germination except in the case of switchgrass, where small seed germinated poorly. Heavy seeds of sand bluestem, sideoats grama, switchgrass, and yellow indiangrass were slightly higher in germination than light seed (Green & Hansen, 1969). Heavy seeds of sideoats grama, blue grama, switchgrass and yellow indiangrass also germinated faster than light seed. Rafii and Barnett (1970) related seed characteristics to field establishment. The seed characteristics evaluated were seed set, caryopsis weight, and germination at various times after harvest. The relationship between spikelet germination and establishment varied with seed age and treatment of the spikelets. They concluded that dormancy was an important aspect of the establishment capabilities of indiangrass. The effects of differential dormancy levels relative to seed age can complicate the determination of establishment capability (Rafii & Barnett, 1970) of seed lots.

The effect of row-spacing and N fertilization on the germination characteristics of seeds produced from stands of 'Blackwell', 'Cave-in-Rock', and 'Pathfinder' switchgrass has been studied (Mullen et al., 1985). Although stand age and differential lodging among cultivars may have influenced seed size and other quality parameters, the data indicate an effect of N fertilization, year of production, and cultivar on dormancy levels. Row spacing did not affect dormancy levels in this study.

Seed production variables appear to influence the final quality and dormancy level in seed. It is the improvement of germination over time, however, that provides an indication that seed dormancy is involved. In general, seed dormancy is prolonged by storage of dry seed at cool temperatures and usually shortened by exposure to higher temperatures (Taylorson & Hendricks, 1977). This is not to be confused with the positive effects of prechill or stratification in promoting the germination of partially or wholly hydrated dormant seed of many species. One factor mediating the seed dormancy response to temperature is seed moisture content. The general storage behavior of dormant seed cited earlier usually holds true for seed that are stored at a low enough seed moisture content that they probably cannot respond physiologically to the temperature effects.

The effects of temperature on germination of several grass species including big bluestem, Caucasian bluestem (*Bothriochloa caucasica* Trin.), indiangrass, and switchgrass have been studied. In one experiment (Hsu et al., 1985) seed were prechilled at 4°C for 2 wk. The germination maximum was between 20 and 30°C for chilled seed versus 12 to 20°C for unchilled seed. Hsu et al. (1985) explained this differential response to temperature via the hypothesis that the cooler temperatures partially fulfilled the prechilling requirement for seed that had not been chilled. Cobb et al. (1961) studied the effects of temperature and hydrogen peroxide on the germination of 'Cobb' switchgrass seed that had been stored 9 to 12 mo. While an alternating temperature of 15/30°C (with or without light or hydrogen peroxide) could induce complete germination in 14 to 28 d, a constant 30°C temperature with light and hydrogen peroxide resulted in complete germination in 7 d (Cobb et al., 1961). The influence of temperature on seed germination in three range grasses was studied in an effort to understand species establishment from seed in the Pawnee National Grassland (Bokhari et al., 1975). The temperature optimum for germination of blue grama and buffalograss in these studies was 29.5/18°C. Bokhari et al. (1975) indicated that moisture conditions in the Pawnee National Grasslands often become dry too quickly for successful establishment events to be common. These problems indicate the need for rapid germination and seedling growth in order for seedling survival in the arid conditions that predominate in many grassland areas.

Germination responses to temperature were studied in Boer lovegrass (*Eragrostis chloromelas* Steud.), Lehmann lovegrass (*E. lehmanniana* Nees), galleta (*Hilaria jamesii* Torr.), and blue grama (Knipe, 1967). The optimum temperature range for germination of blue grama was found to be 60 to 100°F. Knipe (1967), however, removed all attached structures from the caryopses prior to the germination treatments. In general, alternating temperatures were not found to be superior to constant temperatures for germination and Lehmann lovegrass had the most specific temperature requirements for germination. Sosebee and Herbel (1969) investigated the effects of temperature on the germination of 13 different range species. Included in the study were 'Vaughn' and NM-28 sideoats grama. The temperatures used (Sosebee & Herbel, 1969), 18 to 39°C and 18 to 53°C, were chosen to mimic soil temperature environments observed for a soil surface shaded by brush or left bare, respectively. Although the high temperature treatment was detrimental to all species, black grama [*Bouteloua eriopoda* (Torr.) Torr.] and sideoats grama appeared to be better suited to the high temperature regimes. Seeds of freshly harvested 'Coronado' sideoats grama were dormant (Sumner & Cobb, 1962). The duration of dormancy depended on the storage temperature with low temperature (5°C) storage increasing dormancy duration and higher temperatures (20 or 30°C) shortening duration. Sumner and Cob (1962) found that treating seed with sodium hypochlorite effectively overcame dormancy as early as 43 d after harvest. They further reported that the beneficial effect of sodium hypochlorite was due to the removal of inhibitory compounds from the encompassing structures of the seed. This hypothesis was supported by the application of hot water extracts from 90 g of chaff that either had or had not been pretreated with sodium hypochlorite. The first 30 mL of leachate from chaff, which had not been previously treated with sodium hypochlorite, inhibited germination while chaff that had been treated with sodium hypochlorite did not. Washing the seed also improved germination. Two-way thermogradient table studies (Cole et al., 1974) of the temperature requirements for the germination of three sources of sideoats grama indicated that no specific temperature alternation broke dormancy. Cole et al. (1974) reported that specific temperature optima depended on the year of production of the seed. Further, light did not influence seed germination in this study. The absence of a clear temperature optimum in this study is interesting but certainly understandable if dormancy levels among seed sources were variable. As previously described, the requirement for a specific temperature environment declines as dormancy dissipates. This may, in part, explain why the germination temperature ranges of six varieties of 2-yr-old indiangrass seed were similar (Fulbright, 1988).

While it is apparent that dormancy in native warm-season grasses will dissipate in some describable fashion with seed age, marketing demands and other factors often make storage the last choice for controlling seed dormancy. An alternative to storing seed is to find some means to treat seed lots such that rapid germination can be obtained from neoteric seed. Usually this involves either mechanical scarification, chemical modification of dormancy levels, prechilling seed or some combination of these treatments. The following discussions overview research findings related to these treatments.

Complete germination of buffalograss burs was obtained by clipping the distal ends of the burs (Thornton, 1966). Removal of the lemma and palea of the switchgrass floret increased the germination of switchgrass in greenhouse studies (Sautter, 1962). Sautter (1962) also compared the effects of various low temperature treatments, scarification (removal of lemma and palea by rubbing with emery cloth) and various sodium chloride concentrations on seed germination. The scarification treatment induced the highest germination of all treatments, 84% with 74% of the seed germinating in 3 d. Constant low temperatures (38 to 40°F) for 49 d prior to planting resulted in 76% germination. Jensen and Boe (1991) also mechanically scarified seed of switchgrass. Scarification increased the overall mean germination of three lots of 'Sunburst' and two lots of North Dakota Ecotype by 73% (Jensen & Boe, 1991). Coukos (1944) indicated that hammer mill processing or even slight scarification by combining versus hand harvesting was effective in breaking the dormancy in a substantial percentage of caryopses of big and little bluestem. This supports, in part, the work of others who indicate that the mechanical handling and conditioning (processing) of seed can affect both physical and physiological seed quality (Weber et al., 1939; Brown et al., 1983). Coukos (1944) also was able to induce germination by the application of concentrated or dilute acid but observed that these treatments were detrimental to nondormant seed.

The effect of dormancy breaking treatments on laboratory germination and field establishment of 'Holt,' 'Cheyenne,' and 'Osage' indiangrass spikelets has been investigated (Geng & Barnett, 1969). Dormancy breaking treatments included prechilling, hull removal, prechilling followed by hull removal, and hull removal followed by prechilling. Geng and Barnett (1969) removed hulls via a rubbing board and such treatment increased the numbers of abnormal seedlings, which the authors attributed to mechanical injury. The prechill treatment that consisted of exposure of imbibing seeds to 4°C for 2 wk appeared to increase seed germination of Holt but not Cheyenne or Osage. Prechilling resulted in a 16% increase in field establishment of Holt and a nearly significant increase in establishment of Osage seed. This led Geng and Barnett (1969) to question the economic feasibility of prechilling indiangrass seed. They also noted however that prechilling did not appear to remove all of the seed dormancy from indiangrass seed and that further research on dormancy reducing treatments might be warranted. Regression analyses indicated that field establishment was affected by seed set and caryopsis weight in addition to spikelet germination. It was interesting that the field establishment was considerably higher than laboratory germination except in those treatments involving hull removal.

Ahring and Todd (1977) conducted studies on the benefits of the removal of encompassing structures, presoaking burs in sodium hypochlorite and heating (70°C for 12 h) on buffalograss germination. Caryopses, extracted by hand, germinated readily and sodium hypochlorite and heat treatments also improved germination. Heat treatments at 70°C for >12 h reduced seed viability. Additionally, dipping buffalograss caryopses in oils extracted from the burs reduced germination by 47.2%. Dipping root tips into the oil extract resulted in stunted plant growth (Ahring & Todd, 1977).

Prechilling three seed lots of 'KY 1625' switchgrass improved germination by 10 to 26% while soaking seed in 5.25% sodium hypochlorite (bleach) improved germination from 0 to 56% (Panciera et al., 1987). Panciera et al. (1987) found a significant interaction between germination treatments and seed lots. It was further hypothesized that at least two dormancy mechanisms exist in switchgrass.

In a large study involving a total of 13 seed lots of 'Blackwell,' 'Trailblazer,' 'Cave-in-Rock,' 'Kanlow,' and 'KY 1625,' prechilling seed at 5°C for 14 d improved both the speed and quantity of germination (Zarnstorff et al., 1994). Scarification treatments did not consistently improve germination and the effects of the application of hormones (gibberellic acid and kinetin) also were variable. Cultivar and lot influenced the germination response to many of the treatments. Zarnstorff et al. (1994) indicated that postharvest storage of seed at 23°C for 90 d broke dormancy in three seed lots of switchgrass. They recommend further study of the influence of hormones on seed germination. Removal of hulls from blue grama seeds followed by treatment with hormones such as indoleacetic acid reduced the time between germination and adventitious root formation (Roohi & Jameson, 1991). Since the germination rate also was improved, these treatments affected both germination and seedling growth processes. These studies were conducted in order to determine whether such treatments could increase the likelihood of stimulating adventitious root development prior to substantial soil drying in marginal range conditions.

Emal and Conard (1973) studied the influence of light, chilling, sodium hypochlorite, sulfuric acid, and gibberellic acid on the germination of Holt indiangrass seed of various ages. Prechilling seed at 4 to 6°C for 4 wk resulted in nearly complete germination, however, the prechill period could be reduced to 2 wk if the seed were prechilled in light. The quantity and quality of light appeared to influence germination as a 2-h exposure to red light equaled the germination promoting effect of 10 h of daylight. Sodium hypochlorite increased germination if seeds were germinated in daylight but had no effect in dark germinated seeds (Emal & Conard, 1973). Sulfuric acid treatment and the application of gibberellic acid improved germination in both light and dark germinated seedlings although light improved the response to both treatments.

Ahring and Frank (1968) used a South Dakota Blower to separate sterile and fertile cupules of eastern gamagrass (*Tripsacum dactyloides* L.). Following this treatment cupules were exposed to various dormancy reducing treatments. The influence of prechilling (5–10°C), soaking seed in sodium hypochlorite, exposing seed to vapors as well as soaking in various concentrations of ethylene chlorohydrin, and utilizing 0.1 to 0.8 % solutions of KNO_3 as moistening agents for substrate were investigated. Of these treatments only prechilling substantially improved germination of the eastern gamagrass seed lots investigated (Ahring & Frank, 1968). Prechilling for 6 wk appeared to be optimal and differential lot responses to prechilling were noted. Germinations were conducted at 20 to 30°C (16/8 h) and at a constant 30°C with similar germination responses found in each environment. Ahring and Frank (1968) noted heavy mold growth in germination tests except for those involving sodium hypochlorite or ethylene chlorohydrin but indicated that the mold growth did not appear to restrict germination.

Anderson (1985) investigated the germination characteristics of eastern gamagrass seed collected near Carbondale, Murphysboro, and Centralia, Illinois. Removal of the cupule resulted in increased germination rate and quantity. Germination percentages of seeds with cupules removed ranged from 58 to 86% as compared with 10 to 26% germination for seeds with the cupules intact. In one treatment (Anderson, 1985) cupules were removed and the seed then germinated in contact with the cupule. This treatment did not appear to reduce germination. Soaking seed in a gibberellic acid solution (1000 mg mL^{-1}) also improved germination. Germination following a 60-d prechill at 4.4°C was higher when the subsequent germination temperature was 32°C than 25°C. Seed source by incubation treatment interactions were significant.

Soaking seed of three samples of certified 'Pete' eastern gamagrass for 2 h in a 30% hydrogen peroxide solution improved germination an average of 40 to 45% across other treatments that included a 15 min soak in 50% commercial bleach solution and wet-chilled stratified seed (Kindiger, 1994). The wet-chill stratification treatment consisted of a 6 to 9 wk exposure to wet-chilling at 0°C. Germination following these treatments was evaluated after 14 d at 30°C with continuous light. Seeds used in these studies were dehulled.

The possible effects of encompassing structures on dormancy including complete or partial impermeability and the potential presence of inhibitors in these encompassing structures is an important aspect in evaluating practical methods for controlling seed dormancy. This, in conjunction with the known beneficial effects of seed priming have led to several studies evaluating the effects of soaking or priming on the germination of native warm-season grasses.

The germination of buffalograss seed improved with increasing storage for up to 3 to 4 yr (Wenger, 1941). Soaking the burs of buffalograss for 2 to 4 d in tap water increased the germination if immediate and thorough drying followed this soak. The best treatment appeared to be soaking the seed for 72 h that increased germination from 7% in untreated controls to 46.6%. Wenger (1941) indicated that the value of soaking in relation to germination would continue for at least 2 yr. Wenger also noted that soaking the seed in 75% sulfuric acid for 105 min improved germination. Soaking burrs of 'Sharp's Improved' buffalograss burrs in water (Fry et al., 1993), improved establishment parameters such as seedling emergence and coverage. Fry et al. (1993) used different planting dates to determine the optimum date and, while soaking apparently improved the rate of germination, the time to complete foliar coverage of the area planted was not reduced by more than one week, regardless of the time of planting.

Table 7–2. Summary of dormancy breaking treatments reported for certain native warm-season grass species.

Species	"Hull"† removal	Mechanical scarification	Pre-chill	Hor-mones	H_2O_2	Sodium Hypochlorite	Heat-ing	Light	Acid
Big Bluestem		Yes	Yes	Yes‡					
Blue grama	Yes	Yes		Yes					
Buffalograss	Yes	Clipping Burs		Yes‡		Yes	Yes		
Little bluestem		Yes		Yes‡					
Sideoats grama						Yes			
Switchgrass	Yes		Yes	Yes		Yes			
Indiangrass	Yes		Yes	Yes‡				Yes	Yes
Eastern gamagrass	Yes	Yes	Yes		Yes				

† The term "hull", depending on the species, may refer to cupules, burs, or lemma and palea.
‡ Prechill recommended by the Association of Official Seed Analysts, *Rules for Testing Seeds*.

Seed priming has been used to improve germination and establishment of several species (Khan, 1992). Beckman et al., (1993) investigated the effects of solid matrix priming (SMP) on the establishment of big bluestem and switchgrass seed. Seed were either primed for 2 d at 17°C or for 14 d while at 4°C. The latter treatment combined priming with a prechill treatment. Emergence in the greenhouse and field compared these treatments and unprimed controls. Solid matrix priming treatments improved the germination of big bluestem by 18%, however, in the field the emergence of dry untreated seed was greater than or equal to the emergence of SMP-treated seed. Treatments did not influence adventitious root development in big bluestem. Priming for 2 d at 17°C or for 14 d at 4°C improved greenhouse emergence by 35 and 150%, respectively. Solid matrix priming treatments improved emergence in the field when seed were planted in moist soil. Solid matrix priming treatments had the potential to increase stands if seed were not allowed to dry. The effect of seed drying on the value of SMP treatments probably explains the reduced performance of primed seed as compared with untreated seed when planted in dry soil conditions.

SUMMARY

The dormancy behavior of warm-season grass seeds appears to follow general trends for dormancy found in many other grasses. It is apparent that the encompassing structures play an important role in the dormancy of these seed (Table 7–2). The exact nature of this effect is not known at this time. Certainly, the seed coat could partially restrict gas exchange as well as contain inhibitors that reduce germination. Little or no direct evidence has been generated evaluating the role of these encompassing structures on water uptake. Several authors have indicated that more than one dormancy mechanism exists in these seeds but often this evidence is derived from observations that location of production, seed age, and cultivar influence responses to various dormancy breaking treatments with a combination of dormancy breaking treatments improving germination. Certainly, this cannot be taken as proof that more than one dormancy mechanism exists. It does, however, substantiate the need to consider the level of dormancy in a seed lot when attempting to define the behavior of dormancy in given environments or when trying to develop a practical dormancy breaking procedure.

The influence of various dormancy breaking treatments on the germination of native warm-season grasses is sometimes variable. Some of this variation could be due to differential dormancy levels. In the case of the application of exogenous naturally occurring PGRs to seeds, however, the variation in effect could also be due to the methods used for applying these treatments (Palevitch & Thomas, 1976).

A great deal remains to be learned about the germination behavior of seeds of the native warm-season grasses. This work will be important in maximizing the use of these species in agronomic and ecological scenarios.

REFERENCES

Ahring, R.M., and H. Frank. 1968. Establishment of Eastern gamagrass from seed and vegetative propagation. J. Range Manage. 21:27–30.

Ahring, R.M,. and G.W. Todd. 1977. The bur enclosure of the caryopses of buffalograss as a factor affecting germination. Agron. J. 69:15–17.

Amen, R.D. 1968. A model of seed dormancy. Bot. Rev. 34(1):1–29.

Association of Official Seed Analysts. 1993. Rules for testing seeds. J. Seed Technol. 16(3):1–113.

Anderson, R.C. 1985. Aspects of the germination ecology and biomass production of eastern gamagrass (*Tripsacum dactyloides* L.). Bot. Gaz. 146(3):353–364.

Beckman, J.J., L.E. Moser, K. Kubik, and S.S. Waller. 1993. Big bluestem and switchgrass establishment as influenced by seed priming. Agron. J. 85:199–202.

Bewley, J.D., and M. Black. 1982. Physiology and biochemistry of seeds: In relation to germination. Springer-Verlag, New York.

Boe, A., and R. Wynia. 1982. Relationship of age of seed to germination and emergence of switchgrass. Proc. S.D. Acad. Sci. 61:163.

Bokhari, U.G., J.S. Singh, and F.M. Smith. 1975. Influence of temperature regimes and water stress on the germination of three range grasses and its possible ecological significance to a shortgrass prairie. J. Appl. Ecol. 12(1):153–163.

Brown, R.R., J.Henry, and W. Crowder. 1983. Improved processing for high-quality seed of big bluestem, *Andropogon gerardii*, and yellow indiangrass, *Sorghastrum nutans* Livestock forage in the corn belt, USA. p. 272–274. *In* Proc of the XVI Int. Grassland Congress, Lexington, KY. 15–24 June 1981. Westview Press, Boulder, CO.

Chancellor, R.J. 1984. The role of seed dormancy in weed control. Schweiz. Landwirtsch. Forsch. 23:69–75.

Cobb, R.D., D.C. Sumner, and, L.G. Jones. 1961. Some effects of temperature and hydrogen peroxide treatment on the germination of switchgrass (*Panicum virgatum*). Assoc. Off. Seed Anal. Newsl. 35:26.

Cole, D.F., R.L. Major, and L.N. Wright. 1974. Effects of light and temperature on germination of sideoats grama. J. Range Manage. 27(1):41–44.

Coukos, C.J. 1944. Seed dormancy and germination in some native grasses. J. Am. Soc. Agron. 36:337–345.

Crocker, W. 1916. Mechanisms of dormancy in seeds. Am. J. Bot. 3:99–106.

Emal, J.G., and E.C. Conard. 1973. Seed dormancy and germination in Indiangrass as affected by light, chilling, and certain chemical treatments. Agron. J. 65:383–385.

Fry, J., W. Upham, and L. Leuthold. 1993. Seeding month and seed soaking affect buffalograsss establishment. Hort. Sci. 28(9):902–903.

Fulbright, T.E. 1988. Effects of temperature, water potential, and sodium chloride on indiangrass germination. J. Range Manage. 41:207–210.

Geng, S., and F.L. Barnett. 1969. Effects of various dormancy reducing treatments on seed germination and establishment of indiangrass, *Sorghastrum nutans* (L.) Nash. Crop Sci. 9:800–803.

Green, N.E., and R.M. Hansen. 1969. Relationship of seed weight to germination of six grasses. J. Range Manage. 22:133–134.

Heydecker, W. (ed.) 1973. Seed ecology. Pennsylvania State Univ. Press, University Park.

Hsu, F.H., C.J. Nelson, and A.G. Matches. 1985. Temperature effects on germination of perennial warm season forage grasses. Crop Sci. 25:215–220.

Jensen, N.K., and A. Boe. 1991. Germination of mechanically scarified neoteric switchgrass seed. J. Range Manage. 44:299–301.

Karssen, C.M. 1995. Hormonal regulation of seed development, dormancy, and germination studied by genetic control. p. 333–350. *In* J. Kigel and G. Galili (ed.) Seed development and germination. Marcel Dekker, New York.

Kermode, A.R. 1995. Regulatory Mechanisms in the transition from seed development to germination: Interactions between the embryo and the seed environment. p. 273–332. *In* J. Kigel and G. Galili (ed.) Seed development and germination. Marcel Dekker, New York.

Khan, A.A. 1977. The physiology and biochemistry of seed dormancy and germination. North-Holland Publ. Co., Amsterdam.

Khan, A.A. 1982. The physiology and biochemistry of seed development, dormancy, and germination. Elsevier Biomedical Press, New York.

Khan, A.A. 1992. Preplant physiological seed conditioning. Hort. Rev. 13:131–181.

Kindiger, B. 1994. A method to enhance germination of eastern gamagrass. Maydica. 39:53–56.

Kneebone, W.R. 1960. Size of caryopses in buffalograss (*Buchloe dactyloides* (Nutt.) Engelm.) as related to their germination and longevity. Agron. J. 52:553.

Kneebone, W.R., and C.L. Cremer. 1955. The relationship of seed size to seedling vior in some native grass species. Agron. J. 47:472–477.

Knipe, O.D. 1967. Influence of temperature on the germination of some range grasses. J. Range Manage. 20:298–299.

Lang, A. 1965. Effects of some internal and external conditions on seed germination. p. 894–908. *In* W. Ruhland. (ed.) Handbuch der pflanzenphysiologie. Springer-Verlag, Berlin.

Mullen, R.E., P.C. Kassel, and A.D. Knapp. 1985. Seed dormancy and germination of switchgrass from different row spacings and nitrogen levels. J. Appl. Seed Prod. 3:28–33.

Palevitch, D., and T.H. Thomas. 1976. Enhancement by low pH of gibberellin effects on dormant celery seeds and embryoless half-seeds of barley. Physiol. Plant. 37:247–252.

Panciera, M.T., G.A. Jung, and W.C. Sharp. 1987. Switchgrass seedling growth and cultivar dormancy: Potential effects on establishment. p. 244–248. *In* Proc. of the 1987 Grassland Conf. Forages: Resources of the Future, Lexington, KY. 2–5 Mar. Am. For. and Grassland Council, Lexington, KY.

Rafii, Z.E., and F.L. Barnett. 1970. Seed characteristics and field establishment in indiangrass, *Sorghasttrum nutans* (L.) Nash. Crop Sci. 10:258–262.

Roohi, R., and D.A. Jameson. 1991. The effect of hormone, dehulling and seedbed treatments on germination and adventitious root formation in blue grama. J. Range Manage. 44:237–241.

Sautter, E.H. 1962. Germination of switchgrass. J. Range Manage. 15:108–110.

Shaidee, G., B.E. Dahl, and R.M. Hansen. 1969. Germination and emergence of different age seeds of six grasses. J. Range Manage. 22:240–243.

Simpson, G.M. 1990. Seed dormancy in grasses. Cambridge Univ. Press, New York.

Sosebee, R.E., and C. H. Herbel. 1969. Effect of high temperatures on emergence and initial growth of range plants. Agron. J. 61:621–631.

Sumner, D.C., and R. D. Cobb. 1962. Post harvest dormancy of Coronado sideoats Grama *Bouteloua curtipendula* (Michx.) Torr. as affected by storage temperature and germination inhibitors. Crop Sci. 2:321–326.

Taylorson, R.B., and S.B. Hendricks. 1977. Dormancy in seeds. Annu. Rev. Plant Physiol. 28:331–354.

Thornton, M.L. 1966. Seed dormancy in buffalograss (*Buchloe dactyloides*). Proc. Assoc. Off. Seed Anal. 56:120–123.

Vegis, A. 1964. Dormancy in higher plants. Annu. Rev. Plant Physiol. 15:185–224.

Villiers, T.A. 1972. Seed Dormancy. p. 219–281. *In* T.T. Kozlowski (ed.) Seed biology: Germination control, metabolism, and pathology. Academic Press, New York.

Weber, G.L., J.T. Sarvis, and G.A. Rogler. 1939. A method of preparing some native grass seed for handling and seeding. J. Am. Soc. Agron. 31:729.

Wenger, L.E. 1941. Soaking buffalograss (*Buchloe dactyloides*) seed to improve its germination. J. Am. Soc. Agron. 33:135–141.

Zarnstorff, M.E., R.D. Keys, and D.S. Chamblee. 1994. Growth regulator and seed storage effects on switchgrass germination. Agron. J. 86:677–672.

8 Genenetic Advances in Eastern Gamagrass Seed Production

Chet Dewald

USDA-ARS
Woodward, Oklahoma

Bryan Kindiger

USDA-ARS
El Reno, Oklahoma

Eastern gamagrass [*Tripsacum dactyloides* (L.) L.] has long been recognized as a highly productive and palatable forage grass of the eastern prairies (Rechenthin, 1951). Hitchcock and Clothier (1899) attribute the lack of attention given to this robust forage grass partly to its sparse occupation of its native habitat. Characteristics of gamagrass that have limited its abundance and usefulness include low seed production, inferior seed quality, establishment difficulties, and lack of persistence under improper grazing management (Killebrew, 1878; Polk & Adcock, 1964; Ahring & Frank, 1968).

Cutler and Anderson (1941) provide the earliest, and perhaps the most concise, survey of the genus, including taxonomic, distribution, and species classifications. *Tripsacum* comprises 15 species in two sections (Cutler & Anderson, 1941; deWet et al., 1981, 1982, 1983) and is native only to the Western hemisphere. It is widely distributed from Connecticut to Kansas and south to Texas in the USA, south through Mexico and Central America, and deep into Paraguay and Brazil. The genus consists of diploid ($2n = 2x = 36$), triploid ($2n = 3x = 54$), tetraploid ($2n = 4x = 72$), pentaploid ($2n = 5x = 90$) and hexaploid ($2n = 6x = 108$) cytotypes.

Among the earliest cytological and genetic studies of *Tripsacum* were those of Mangelsdorf and Reeves (1931, 1939), Anderson (1944), Maguire (1952), Prywer (1954), and Tantravahi (1968). It appears, however, Farquharson (1954, 1955) was the first to attempt a characterization of the reproductive processes found in *Tripsacum*. Subsequent studies employing embryological techniques indicate that diploids reproduce exclusively by sexual means while polyploid cytotypes reproduce as facultative apomicts. Apomixis reproduction in *Tripsacum* is characterized as being diplosporous pseudogamy of the *Antennaria* type (Burson et al., 1990; Sherman et al., 1991; Leblanc et al., 1995). The facultative nature of apomixis in eastern gamagrass, however, is questioned from results obtained during progeny tests of triploid and tetraploid apomictic forms (Kindiger & Dewald, 1994, 1996).

Copyright © 2000. Crop Science Society of America and American Society of Agronomy, 677 S. Segoe Rd., Madison, WI 53711, USA. *Native Warm-Season Grasses: Research Trends and Issues*. CSSA Special Publication no. 30.

CLASSICAL MONOECIOUS BREEDING SYSTEM

The classical *Tripsacum* inflorescence is monoecious with separation of the sexes on the same inflorescence (Conner, 1979). Hitchcock and Chase (1951) described *Tripsacum* of the USA as having unisexual spikelets, with solitary female spikelets containing one fertile floret below, and paired male spikelets, each with two fertile florets above on a continuous rachis (Fig. 8–1). Terminal infloresences of eastern gamagrass typically have 1 to 5 racemes per inflorescence with the seed bearing portion being one-fourth or less of the total (14–25 cm) inflorescence length. At Woodward, OK, the ratio of female flowers to anthers is 1:54, further evidence of the excessive tendency towards maleness prevalent in the *Andropogoneae* tribe to which *Tripsacum* belongs (Dewald & Jackson, 1986).

PROLIFIC GYNOMONOECIOUS BREEDING SYSTEM

In the Poaceae, atavistic reversal or rudimentary organs recovering their original functions, is fairly common and has been given the term *genetic recall* by Harlan (1981). We concur with this hypothesis that alien genomes combined in the same nuclei will influence regulatory functions and affect morphogenesis in numerous ways. An example of regulatory action on morphogenesis is the gynomo-

Fig. 8–1. The classical monoecious inflorescence of *Tripsacum* with solitary female spikelets containing one fertile floret below, and paired male spikelets, each with two fertile florets above on a continuous rachis. Terminal infloresences of *T. dactyloides* typically have 1 to 5 racemes per inflorescence with the seed bearing portion being one-quarter or less of the total (15 to 25 cm) inflorescence length. Extreme maleness and a scarcity of female florets severely limits the seed production potential of *Tripsacum*.

noecious sex form of gamagrass, *T. dactyloides* var. *prolificum*, which exhibits a gynomonoecious breeding system characterized by female flowers below and bisexual flowers above on the same raceme (Fig. 8–2). This variant has the potential to produce two seeds per spikelet instead of one, and has both female and bisexual florets rather than male spikelets in the terminal (tassel) portion of the inflorescence. Basal pistillate spikelets (4–18) are subsessile, usually solitary, and seeds (usually two) are enclosed in cupulate fruitcases. Above the cupulate fruitcase association, the inflorescence structure resembles the tassel or staminate portion of the normal sex form except the outer glumes are indurate and wider, having female secondary sex characteristics (Dewald & Dayton, 1985a,b). This novel variant is fully fertile and produces viable seed throughout the inflorescence when pollinated by a wide range of normal diploid *T. dactyloides* strains. The gynomonoecious variant's prolific seed production potential (20- to 25-fold increase) and the production of seed in the tassel, free of the cupulate fruitcase and therefore threshable, suggests potential for development of a perennial grain crop. This variant has been used extensively in our breeding program to produce about 125 000 hybrids, many with additional traits that will enhance grain production in *Tripsacum* (Dewald & Sims, 1990). This prolific seed-producing variant condition is regulated by a recessive major gene ($gsf1/gsf1$ = gynomonoecious sex form) at a single locus (Dewald et al., 1987).

Fig. 8–2. A prolific sex form variant of eastern gamagrass (*T. dactyloides*), which exhibits a gynomonoecious breeding system and is characterized by female flowers below and bisexual flowers above on the same raceme. It has the potential to produce two seeds per spikelet instead of one, and has both female and bisexual florets rather than male spikelets in the terminal (tassel) portion of the inflorescence. Seed production is enhanced because viable seeds can be produced throughout the inflorescence. This characteristic provides for 20 to 25 times more female florets and a subsequent increase in seed yield.

HORIZONTAL GENE FLOW

In this genus, crosses performed among cytotypes possessing varying ploidies ($2x$, $3x$, $4x$, $5x$, and $6x$) and between species are generally accomplished with ease and provide a myriad of new germplasm for evaluation. Crosses between sexual diploids and apomictic tetraploids generally result in both sterile and fertile triploid hybrids (Dewald et al., 1992). Sterile triploids are believed to reproduce by sexual means with the imbalance in their chromosome number causing frequent abortion of pollen and the developing megaspore. Fertile triploids reproduce exclusively by apomixis, but also can produce hybrids via $2n + n$ matings. Such hybrids, commonly called B_{III} derived hybrids (Bashaw et al., 1992), are generated by fertilization of an unreduced egg with the paternal genome contribution from a pollen parent (Kindiger & Dewald, 1994). A fertile, apomictic triploid × sexual diploid cross often results in a new tetraploid hybrid. By using this unique attribute, movement of traits between sexual and apomictic genotypes can be accomplished readily as exemplified by the development of fertile, gynomonoecious (GSF) triploids (Dewald & Kindiger, 1994, 1996). In addition, by using this behavior, tetraploid × diploid and tetraploid × tetraploid crosses can be used to generate both pentaploid and hexaploid genotypes. Our research on pentaploids and hexaploids have shown them to be 99 to 100% female sterile, presumably due to chromosome imbalance. Hexaploids, however, often are male fertile and can be used to generate, upon crossing to diploids, individuals with aneuploid, diploid, triploid, and tetraploid genotypes (Kindiger & Dewald, 1998, unpublished data). In addition, many hexaploid individuals ($2n = 6x = 108$) often exhibit 100% maternal sterility, although some hexaploids exhibit a good level of apomictic maternal fertility (Kindiger & Dewald, 1998, unpublished data). This reproductive attribute provides ready access to tetraploid germplasm and the manipulation and transfer of genes among and within ploidies (Fig. 8–3). This behavior also has occurred in interspecific crosses between *Tripsacum* species.

GENETIC ALTERATIONS WITHOUT HYBRIDIZATION

All fertile polyploid eastern gamagrass reproduce by apomixis; however, PCR-RAPD generated profiles have identified slight variations in the genetic pro-

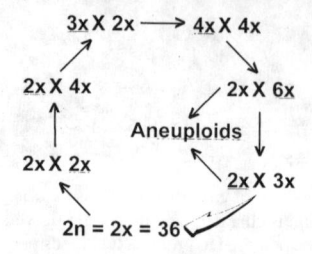

Fig. 8–3. Flowchart depicting the various reproductive pathways observed in eastern gamagrass. The diagram presents the major products generated from the various crossing schemes. Individuals highlighted with bold italicized numbers and underlined indicate the new genotypes generated from the specific cross.

files of gamagrass individuals derived by apomixis, suggesting a level of genetic segregation–recombination was taking place with no reduction in chromosome number or contribution from outside pollen sources. Koltunow (1993) previously described a process whereby variant or off-type genotypes could be generated in plants exhibiting mitotic diplospory. This change would arise through an incomplete meiotic event, altering the genetic constitution of an individual but retaining the original chromosome number. A study of two different families consisting of 100 individuals each derived from apomictic tetraploid parents of *T. dactyloides* indicated the possible occurrence of a first division restitution (FDR), which resulted in altered genotypes with no alteration in chromosome number. This system allows for genetic alterations in a previously suspect closed apomictic reproductive system (Kindiger & Dewald, 1996a).

BREEDING STRATEGIES

The characterization of gamagrass accessions for agronomic traits and their subsequent use in the development of superior cultivars is both feasible and practical (Vogel et al., 1985). Elucidation of the various reproductive mechanisms found in apomictic eastern gamagrass has allowed us to develop several scenarios for breeding strategies directed toward commercial seed production. Two that appear quite useful rely on development of apomictic hybrids. Hybrid apomictic varieties would provide agronomic uniformity, predictability of harvest and grazing periods, and an infinite succession of generations that would perpetuate the original level of hybrid vigor. Since gynomonoecious (GSF) plants are very poor pollen producers, an ample pollen source must be available. Two possible genotypes are available that would provide ample pollen but not produce seed. These are female sterile, hexaploid, and tetraploid germplasm. As discussed earlier, hexaploid germplasm is generated readily by intercrossing tetraploid genotypes and selecting for B_{III} derived individual $(2n + n)$ matings. Female sterile tetraploid germplasm is generated readily by a similar method. In this case, highly sterile triploids are backcrossed by sexual diploids. All B_{III} derived tetraploids generated from this crossing scheme (near sterile $3X \times 2X$) have been found to be female sterile, presumably due to a loss of apomictic alleles (Kindiger & Dewald, 1998, unpublished data). In both cases, B_{III} derived individuals can be identified most easily by chromosome counts or phenotypic examination.

Using the previous genotypes, one feasible production strategy would establish a seed production nursery where pollen deficient, GSF triploid females would be pollinated by female sterile pollen sources. In this scenario, an agronomically superior, GSF, apomictic, triploid hybrid would be pollinated by either female sterile, tetraploid or hexaploid genotypes. This system is identical to the top-cross block style (4 rows female to 2 rows pollinator) used in commercial hybrid maize (*Zea mays* L.) and sorghum [*Sorghum bicolor* (L.) Moench] industries. The GSF seed production rows would produce apomictically derived seed, genetically identical to the maternal parent. The tetraploid or hexaploid pollinators would supply adequate pollen for fertilization, but themselves produce no seed. The inability of the pollinators to produce seed eliminates the problem of dropped seed (field contamination) in the perennial production nursery. Lost seed from the female rows is of no consequence since the seed will be apomictically derived or a product of $2n + n$ matings.

A second possible production situation would be to use sexual diploid, GSF females in crosses with female sterile tetraploids. In this crossing scheme all seed would be F_1, triploid hybrids ($2X \times 4X = 3X$). Once planted, all seed would produce non-GSF hybrids. Due to their sexual reproductive mode and chromosome imbalance at the triploid level, essentially all the hybrids would be male or female sterile or both. Such triploid hybrids should provide excellent forage but possess no value for seed production purposes.

Both of these methods produce a superior product, with known genetic identity, that is controlled by the seed producer. In time, a commercial grower could develop and market protected varieties possessing particular agronomic attributes favorable to their particular markets. It is hoped that in the future, such production systems will be initiated with subsequent acceptance of eastern gamagrass as a viable commercial seed crop.

COLLINEARITY BETWEEN EASTERN GAMAGRASS AND MAIZE

Eastern gamagrass is distantly related to maize, which is the most efficient grain producer of all the grasses (Mangelsdorf, 1986) and the world's most important grain crop. Unimproved *Tripsacum* is a notoriously poor seed producer with yields normally below 100 kg ha^{-1}. In the 1930s, attempts to understand the evolutionary origin of maize were initiated through wide-hybridization experiments with maize and *Tripsacum* (Mangelsdorf & Reeves, 1931, 1939). These early experiments spawned numerous research studies relating to evolution taxonomy and the improvement of maize, which in turn made various contributions toward the study of maize × *Tripsacum* wide hybridization (Maguire, 1961, 1962, 1963; Reeves & Bockholt, 1964; Newell & deWet, 1974; Simone & Hooker, 1976; Harlan & deWet, 1977; Petrov et al., 1979; James, 1979; Cohen & Galinat, 1984; Bergquist, 1981).

Among the Poaceae, recent molecular studies have identified a conservation of linkage groups (Bennetzen & Freeling, 1993). Included are the best known cereals, rice (*Oryza sativa* L.), wheat (*Triticum aestivum* L.), and maize, which account for about one-half of total world food production (Whitkus et al., 1992; Ahn & Tanksley, 1993). Recently, it has been shown that *Tripsacum* also can be included in this category (Blakey, 1993). Comparative mapping strategies that capitalize on potential gene order or conservation among the Poaceae could prove valuable for identifying regions possessing alleles or genes conferring favorable agronomic traits (Hake & Walbot, 1980). Recently, Blakey et al. (1994) identified a conserved group of RFLP markers and possible associated genes on the short arm of chromosome 1 of maize and linkage group I of *Tripsacum*. The *gsf1* allele, which confers sex reversal in the male portion of the *Tripsacum* inflorescence (Dewald et al., 1987), has been assigned to linkage group I. In maize, a tassel seed mutant (*ts2*) has been assigned to the short arm of chromosome 1. Based upon this possible conservation of linkage groups, a test of allelism was performed between a maize line carrying the *ts2* allele and a *gsf1* stock of *T. dactyloides*. A successful test of allelic noncomplementation suggested that the two genes were identical and express identical levels of genetic function and regulation (Kindiger et al., 1994). Additional studies also were successful in identifying regions of the maize and *Tripsacum* genomes with similar genetic structure (Rao & Galinat, 1974, 1976; Kindiger & Beckett, 1989). These results strongly indicate the possibility of using data available from the maize genetic database to facilitate improvements in a *Tripsacum* or maize–*Tripsacum* breeding or genetics program. Recent isozyme, PCR-RAPD, and RFLP investigations indicate that these techniques can be applied to a maize–*Tripsacum* hybridization study where identifying the introgressed *Tripsacum* germplasm in a maize background is necessary (Kindiger & Vierling, 1994; Kindiger et al., 1996b).

CULTURAL ADVANCES

Recent cultural advances in seed production are primarily concerned with agronomic practices that favor the growth of the crop, eastern gamagrass, through reduction of competition (weeds, insects, and others) and improved fertility practices. Seed production problems will be different from area to area and are best handled on a local level. A few general comments may be useful at this time.

Row Spacing

Plantings in wide (36–48″) rows will produce more seed than narrow row or solid plantings and will facilitate travel over the field for ease of cultivation, pesticide application, seed harvesting, and more.

Soil Fertility

Frequent soil testing is necessary to determine proper fertilizer needs and application rates. Soil P level should be maintained on the high side and an annual application of N will be needed in most areas.

Timing of Harvest

Eastern gamagrass seed maturity is indeterminate with seed shattering from the inflorescence as they mature. Timing of harvest is therefore judgmental and should be based on frequent field inspections to determine caryopses content or grain fill of the cupulate fruitcase exclosure. In most cases some seed will have been lost by shattering prior to ideal timing of harvest whereas other seed will be immature. The object of a good seed harvest is to obtain the maximum amount of well filled grain as possible. This is subjective at best, however, seed producers with a little experience seem to be rather good at selecting the near optimum timing of harvest.

Female florets of eastern gamagrass reach sexual maturity a few to several days prior to male florets on the same inflorescence. This leads to seed production failure of the earlier produced inflorescence due to a lack of pollen for fertilization. Heads produced later will be cross pollinated with pollen from earlier inflorescences resulting in much improved seed set. Also, secondary or axillary inflorescences reach sexual maturity later than primary or terminal inflorescences and usually have a higher percentage seed set especially on earlier flowering seed stalks. For these reasons, more mistakes in timing of harvest occur by being too early rather than too late.

SUMMARY

1. The classical monoecious breeding system of eastern gamagrass is a rather inefficient system for seed production due to the scarcity of female seed bearing structures.
2. The prolific gynomonoecious breeding system results from sex reversal, male to female, in the upper (formerly tassel) section of each raceme and reactivation of suppressed female florets in cupulate fruitcases at the base of the raceme. This prolific seed-producing variant condition is regulated by a recessive major gene($gsf1/gsf1$ gynomonoecious sex form) at a single locus. This characteristic provides for 20 to 25 times more female florets and subsequent increase in seed yield.
3. Diploid ($2n = 2x = 36$) cytotypes of eastern gamagrass reproduce sexually allowing genetic recombination and segregation between lines, races, and even species. This facilitates genetic diversity and trait intensification through intermating, back crossing and selfing schemes.
4. Female fertile polyploids ($3x$, $4x$, $5x$, and $6x$) reproduce primarily through apomixis (diplosporous pseudogamy) and are useful for gene exchange through male gametes and for stabilization of true breeding lines.
5. Horizontal gene flow among *Tripsacum* cytotypes ($2x$, $3x$, $4x$, $5x$, and $6x$) and between species can provide a myriad of new germplasm combinations for evaluation. Crosses between diploid and tetraploid cytotypes usually result in the production of triploids, some (5–10%) of which are female fertile through apomictic reproduction.
6. Genome accumulation through fertilization of unreduced egg cells results in the production of B_{III} *Tripsacum* hybrids with the paternal contribution from a pollen parent. A fertile apomictic triploid × sexual diploid cross often results in new tetraploid hybrids. By using this unique attribute, movement of traits between sexual and apomictic genotypes can be readily accomplished.
7. The reproductive versatility of *Tripsacum* is dynamic and will facilitate development of superior seed producing genotypes and stabilization of true breeding cultivars through apomixis.

REFERENCES

Ahring, R.M., and H. Frank. 1968. Establishment of eastern gamagrass from seed and vegetative propagation. J. Range Manage. 21:27–30.

Ahn, S., and S.D. Tanksley. 1993. Comparative linkage maps of the rice and maize genomes. PNAS USA 90:7980–7984.

Anderson, E. 1944. Cytological observations on *Tripsacum dactyloides*. Ann. Missouri Bot. Garden. 31:317–323.
Bashaw, E.C., M.A. Hussey, and K.W. Hignight. 1992. Hybridization (N+N and 2N+N) of facultative apomictic species in the *Pennisetum* agamic complex. Int. J. Plant Sci. 153:466–470.
Bennetzen, J., and M. Freeling. 1993. Grasses as a single genetic system: genome composition, collinearity and compatibility. Trends Genet. 9:259–292.
Bergquist, R.R. 1981. Transfer from *Tripsacum dactyloides* to corn of a major gene locus conditioning resistance to *Puccinia sorghi*. Phytopathology 71:518–520.
Blakey, C.A. 1993. A molecular map in *Tripsacum dactyloides*, Eastern gamagrass. Ph.D. diss. Univ. Missouri, Columbia (no. 9412465).
Blakey, C.A., C.L. Dewald, and E.H. Coe. 1994. Co-segregation of the gynomonoecious sex form- I *(gsfl)* gene of *Tripsacum dactyloides* (Poaceae) with molecular markers. Genome 37:809–812.
Brown, W.V., and W.H.P. Emery. 1958. Apomixis in the Gramineae: Panicoideae. Am. J. Bot. 45:253–263.
Burson, B.L., P.W. Voigt, R.A. Sherman, and C.L. Dewald. 1990. Apomixis and sexuality in eastern gamagrass. Crop Sci. 30:86–89.
Cohen, J.L., and W.C. Galinat. 1984. Potential use of alien germplasm for maize improvement. Crop Sci. 24:1011–1015.
Connor, H.E. 1979. Breeding systems in the grasses: A survey. NZ J. of Botany 17:547–574.
Cutler, H.C., and E. Anderson. 1941. Preliminary survey of the genus *Tripsacum*. Ann. Missouri Bot. Garden. 28:249–269.
Dewald, C.L., and R.S. Dayton. 1985a. A prolific sex form variant of eastern gamagrass. Phytologia 57:156.
Dewald, C.L., and R.S. Dayton. 1985b. Registration of gynomonoecious germplasm (GSF-1 and GSF-11) of Eastern gamagrass. Crop Sci. 25:715.
Dewald, C.L., and L. Jackson. 1986. Breeding system influence on spikelet number, glume dimensions, and sex ratios of eastern gamagrass. p. 04–05.05. *In* A. Davis and G. Standford (ed.) Proc. 10th North Am. Prairie Conf., Denton, TX. June, 1986. Native Prairie Association of Texas, Dallas.
Dewald, C.L., B.L. Burson, J.M.J. deWet, and J.R. Harlan. 1987. Morphology, inheritance and evolutionary significance of sex reversal in *Tripsacum dactyloides* (Poaceae). Am. J. Bot. 74:1055–1059.
Dewald, C.L., and P.L. Sims. 1990. Breeding *Tripsacum* for increased grain production. p. 27–30. *In* Proc. Eastern Gamagrass Conf. Poteau, OK. 23–25 Jan. 1990. Kerr Center for Sustainable Agriculture, Poteau, OK.
Dewald, C.L., C.M. Taliaferro, and P.C. Dunfield. 1992. Registration of four fertile triploid germplasm lines of eastern gamagrass. Crop Sci. 32:504.
Dewald, C.L., and B. Kindiger. 1994. Genetic transfer of gynomonecy from diploid to triploid eastern gamagrass. Crop Sci. 34:1259–1262.
Dewald, C.L., and B. Kindiger. 1996. Registration of FGT-1 Eastern gamagrass germplasm.Crop Sci. 36:219.
deWet, J.M.J., D.H. Timothy, K.W. Hilu, and G.B. Fletcher. 1981. Systematics of South American *Tripsacum* (Gramineae). Am. J. Bot. 68:269–276.
deWet, J.M.J., J.R. Harlan, and D.E. Brink. 1982. Systematics of *Tripsacum dactyloides* (Gramineae). Am. J. Bot. 69:1251–1257.
deWet, J.M.J., D.E. Brink, and C.E. Cohen. 1983. Systematics of *Tripsacum section* Fasciculata (Gramineae). Am. J. Bot. 70:1139–1146.
Farquharson, L.I. 1954. Natural selection of tetraploids in a mixed colony of *Tripsacum dactyloides*. Proc. Indiana Acad. Sci. 63:80–82.
Farquharson, L.I. 1955. Apomixis and polyembryony in *Tripsacum dactyloides*. Am. J. Bot. 42:737–743.
Hake, S., and V. Walbot. 1980. The genome of *Zea mays*, its organization and homology to related grasses. Chromosoma 79:251–270.
Harlan, J.R., and J.M.J. deWet. 1977. Pathways of genetic transfer from *Tripsacum to Zea mays*. Proc. Natl. Acad. Sci. USA. 74:3494–3497.
Harlan, J.R. 1981. Directing the accelerated evolution of crop plants. p. 51–69. *In* Strategies of plant reproduction. BARC Symp.,. Totawa, NJ. 17–29 May 1981. Allanheld, Osmum & Company, Totawa, NJ.
Hitchcock, A.S., and G.L. Clothier. 1899. Native agricultural grasses of Kansas. p. 5–6. *In* Kansas State Agric. Coll. Bull. 87. Univ. of Kansas, Manhattan.
Hitchcock, A.S., and A. Chase. 1951. Manual of the grasses of the United States. U.S. Gov. Print. Office, Washington, DC.

James, J. 1979. New maize × *Tripsacum* hybrids for maize improvement. Euphytica 28:239–247.
Killebrew, J.B. 1878. The grasses of Tennessee including cereals and forage plants. The American Co., Nashville.
Kindiger, B., and J.B. Beckett. 1989. Cytological evidence supporting a procedure for directing and enhancing pairing between maize and *Tripsacum*. Genome 33:495–500.
Kindiger, B., and C.L. Dewald. 1994. Genome accumulation in eastern gamagrass, *Tripsacum dactyloides* (L.) L. (Poaceae). Genetica 92:197–201.
Kindiger, B., and R. Vierling. 1994. Comparative isozyme polymorphisms of north American eastern gamagrass, *Tripsacum dactyloides var. dactyloides* and maize, *Zea mays L*. Genetica 94:77–83.
Kindiger, B., C.A. Blakey, and C.L. Dewald. 1994. Sex reversal in maize × *Tripsacum* hybrids: Allelic noncomplementation of *ts2 and gsf1*. Maydica 40:187–190.
Kindiger, B., and C.L. Dewald. 1996a. A system for genetic change in apomictic eastern gamagrass. Crop Sci. 36:250–255.
Kindiger, B., V. Sokolov, and C.L. Dewald. 1996b. A comparison of apomictic reproduction in eastern gamagrass [*Tripsacum dactyloides* (L) L.] and *maize–Tripsacum* hybrids. Genetica 97:103–110.
Koltunow, A.M. 1993. Apomixis: Embryo sacs and embryos formed without meiosis or fertilization in ovules. The Plant Cell 5:1425–1437.
Leblanc, O., M.D. Peel, J.G. Carman, and Y. Savidan. 1995. Megasporogenesis and megagametogenesis in several *Tripsacum* species (Poaceae). Am. J. Bot. 82:57–63.
Maguire, M.P. 1952. Synapsis and crossing over of *Tripsacum and* Zea chromosomes derived from hybrids backcrossed to *Zea*. Ph.D. diss. Cornell University, Ithaca, NY.
Maguire, M.P. 1961. Divergence in *Tripsacum* and *Zea* chromosomes. Evolution 15:394–400.
Maguire, M.P. 1962. Chromatid interchange in allodiploid maize-Tripsacum hybrids. Can. J. Gent. Cytol. 5:414–420.
Maguire, M.P. 1963. Crossing over and anaphase I distribution of the chromosomes of a maize interchange trivalent. Genetics 49:69–80.
Mangelsdorf, P.C., and R.G. Reeves. 1931. Hybridization of maize, *Tripsacum* and *Euchlaena*. J. Hered. 22:329–343.
Mangelsdorf, P.C., and R.G. Reeves. 1939. The origin of Indian corn and its relatives. Texas Agric. Exp. Stn. Bull. 574. Texas A&M Univ., College Station.
Mangelsdorf, P.C. 1986. The origin of corn. Sci. Am. 255:80–87.
Newell, C.A., and J.M.J. deWet. 1974. Morphology of some *maize–Tripsacum* hybrids. Am. J. Bot. 61:45–53.
Petrov, D.F., N.I. Belousova, and E.S. Fokina. 1979. Inheritance of apomixis and its elements in maize × *Tripsacum dactyloides* hybrids. Genetika 15:1827–1836.
Polk, D.B., and W.L. Adcock. 1964. Eastern gamagrass. Cattleman 50:83–84.
Prywer, C. 1954. Meiosis en *Tripsacum maizar*. Revista de la Soc. Mexicana de Hist. Nat. Tomo XV. no. 1–4:59–64.
Rao, B.G.S., and W.C. Galinat. 1974. The evolution of the American Maydeae: I. The characteristics of two *Tripsacum* chromosomes (*Tr7* and *Tr 13*) that are partial homoeologs to maize chromosome 4. J. Hered. 65:335–340.
Rao, B.G.S., and W.C. Galinat. 1976. The evolution of the American Maydeae: II. The characteristics of a *Tripsacum* chromosome (*Tr9*) homoeologous to maize chromosome 2. J. Hered. 67:235–240.
Rechenthin, C.A. 1951. Range grasses in the Southwest; eastern gamagrass, Texas cupgrass, Pan American balsomscale and smooth cordgrass. Cattleman 38:110–112.
Reeves, R.G., and A.J. Bockholt. 1964. Modification and improvement of a maize inbred by crossing it with *Tripsacum*. Crop Sci. 4:7–10.
Sherman, R.A., P.W. Voigt, B.L. Burson, and C.L. Dewald. 1991. Apomixis in diploid × triploid *Tripsacum dactyloides* hybrids. Genome 34:528–532.
Simone, G.W., and A.L. Hooker. 1976. Monogenic resistance in corn to *Helminthosporium turcicum* derived from *Tripsacum floridanum*. Proc. Am. Phytopathol. Soc. 3:207.
Tantravahi, R.V. 1968. Cytology and crossability relationships of *Tripsacum*. Ph.D. diss. The Bussey Inst., Harvard Univ. Publication, Cambridge.
Vogel, K.P., C.L. Dewald, H.J. Gorz, and F.A. Haskins. 1985. Improvements of switchgrass, indiangrass, and eastern gamagrass: Current status and future. p. 159–170. *In* Proc. 38th Annual Soc. for Range Manage. Meeting, Salt Lake City, UT. 19 Feb. 1985. Soc. Range Manage., Denver, CO.
Whitkus, R., J. Doebley, and M. Lee. 1992. Comparative genome mapping of sorghum and maize. Genetics 132:1119–1130.

9 Cutting Management of Native Warm-Season Perennial Grasses: Morphological and Physiological Responses

Matt A. Sanderson

USDA-ARS Pasture Systems and Watershed Management Research Laboratory
University Park, Pennsylvania

One of the unique traits of the native perennial grasses is that they were naturally selected for persistence in the face of biotic and abiotic stresses characteristic of North American ecosystems. This does not imply, however, that these grasses are invulnerable. Indeed, a stress imposed at a vulnerable time, at a great enough severity, for a sufficiently long period will inevitably cause plant death. For example, clipping too closely, too frequently, and at a vulnerable stage of development will stress and weaken stands of warm-season grasses (Weaver & Hougen, 1939; Jameson, 1963).

Harvest management of native grasses most often focuses on forage production for livestock. Warm-season native grasses, however, may be used as biofuel crops in the future (McLaughlin, 1993; Sanderson, et al., 1996; Vogel, 1996). Harvest management for biomass production may be different than for hay production because the objective is to obtain high yields of lignocellulose and digestibility is not a consideration. Thus, a single late-season harvest may work best for biomass fuel cropping. Producers, however, may want to integrate forage and biomass cropping for flexibility and diversity in the farming operation.

This chapter focuses on the cutting management (machine harvest) of native warm-season perennial grasses and how the morphology and physiology of the plant dictate appropriate machine harvest systems. The response of grasses to herbivory is discussed in another chapter in this publication (Anderson, 2000, this publication). Most of the discussion centers on the species from the tall grass prairie, viz., switchgrass (*Panicum virgatum* L.), big bluestem (*Andropogon gerardii* Vitman), indiangrass [*Sorghastrum nutans* (L.) Nash], and eastern gamagrass [*Tripsacum dactyloides* (L.) L.] because these are most often machine harvested for conserved forage.

Copyright © 2000. Crop Science Society of America and American Society of Agronomy, 677 S. Segoe Rd., Madison, WI 53711, USA. *Native Warm-Season Grasses: Research Trends and Issues.* CSSA Special Publication no. 30.

MORPHOLOGICAL AND PHYSIOLOGICAL RESPONSES TO CUTTING MANAGEMENT

Developmental Morphology

Cutting management of native perennial warm-season grasses requires an understanding of their developmental morphology and response to defoliation. Developmental morphology is addressed in more detail elsewhere in this publication (Mitchell & Moser, 2000, this publication), thus, only a brief discussion occurs here. The basic sequence of events in the development of most native grasses during spring to fall is a period of leaf production in the early spring, followed by internode elongation and elevation of the apical meristem, transition of the apical meristem from vegetative to reproductive (double ridge formation), inflorescence emergence, flowering, and seed set (Hyder, 1974). Depending on the grass species, the developmental events may be predictable by mathematical models (Sanderson, 1992; Sanderson & Wolf, 1995a,b; Moore & Moser, 1995; Frank & Bauer, 1995).

The location and activity of meristems in the grass canopy determines the physiological and morphological responses to defoliation (Nelson, 1994). New leaves produced in the spring arise from leaf primordia on meristems at or near ground level. Thus, clipping at this time removes mainly leaf blades with some sheath and the intact meristem continues to produce new leaf primordia and leaves. During internode elongation (jointing) new growth from intercalary meristems at the nodes and internodes below the apical meristem causes the culm to elongate and elevates the apical meristem. Depending on clipping height, harvesting at this developmental stage may remove the apical meristem, which stops new leaf development on that particular tiller. Regrowth must then occur from meristems in the axils of previous leaves on the stubble, basal buds on the crown, or rhizomes. If the apical meristem is not removed by clipping, the tiller will continue to elongate and produce new leaves until the meristem changes from vegetative to reproductive. Most native perennial grasses are harvested for hay near the end of the internode elongation phase, at or just before the boot stage (Moser & Vogel, 1994). Thus, the apical meristem will be removed from nearly all tillers by a harvest at this developmental stage. Grass species differ in the timing of apical meristem elevation. Switchgrass typically elevates the apical meristem earlier in the season than big bluestem and indiangrass (Fig. 9–1; Gerrish et al., 1987; Moser & Vogel, 1994). Timing of apical meristem elevation also differs among cultivars within a species, and among environments (Fig. 9–2).

Changes in light quality result in photomorphogenic changes in forages. The plant canopy not only intercepts photosynthetically active radiation (400 to 700 nm wavelengths), but also selectively absorbs red wavelengths (approximately 660 nm) more than far-red wavelengths (approximately 730 nm) such that the red/far-red ratio of light decreases toward the base of the canopy (Holmes & Smith, 1977). Common responses of forage plants to altered light quality include elongation of petioles, leaves, and stems, and reduced tillering in grasses and branching in legumes (Balleré et al., 1995). Frank and Hofmann (1994) demonstrated that defoliation management of perennial grasses affects the red/far-red light ratio in the canopy and stem density. Management that resulted in increased standing forage (e.g., exclu-

CUTTING MANAGEMENT

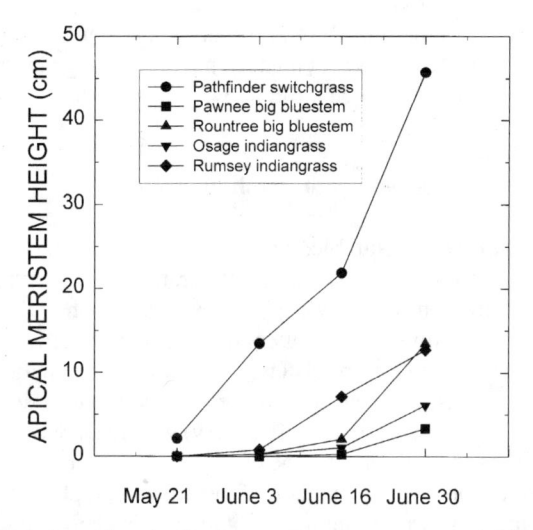

Fig. 9–1. Timing of apical meristem elevation in several warm-season grasses in Missouri. Data are from Gerrish et al. (1987).

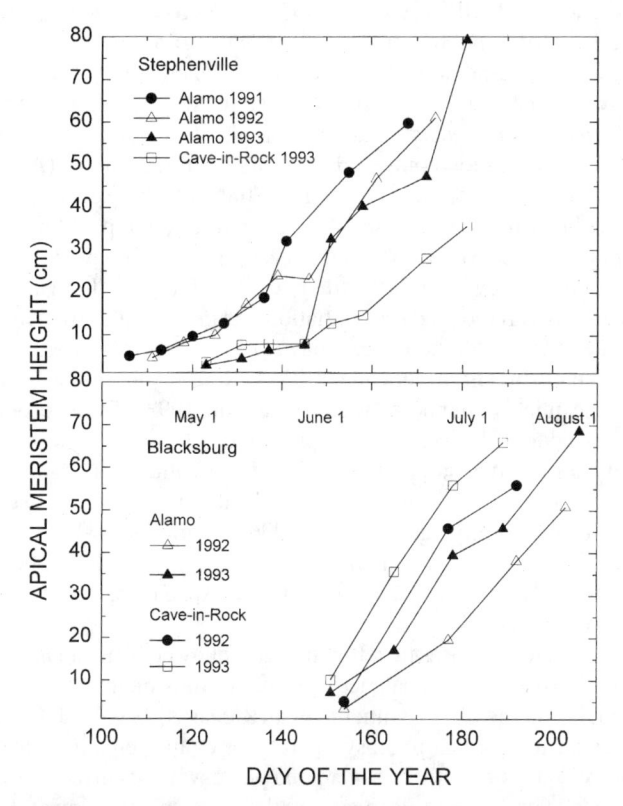

Fig. 9–2. Elevation of the apical meristem in switchgrass during several years at two locations. Data are from Sanderson and Wolf (1995a).

sion of grazing) reduced the red/far-red ratio at the base of the plant canopy and reduced the number of stems per unit area. Removal of forage by grazing or haying increased the red/far-red ratio at the canopy base and increased stem density in the stand.

Responses to Defoliation Stress

Defoliation stresses the stubble, roots, and rhizomes of grasses by depriving them partially or totally of synthesized carbohydrate, whereas respiration continues. The degree of stress imposed by clipping a warm-season grass depends on (i) how closely the plant is clipped (intensity of defoliation), (ii) the physiological age and type of tissue removed (e.g., meristematic vs. nonmeristematic), (iii) clipping frequency, whether in discrete well spaced events or continuous removal, (iv) timing of defoliation during the growth cycle, and (v) whether stresses or competition have occurred before, during, or after the defoliation (Richards, 1993). Photosynthesis of the grass canopy is reduced more if mostly young, photosynthetically active leaves are removed by clipping than if older, shaded leaves of a lower photosynthetic capacity are removed (Gold & Caldwell, 1989a,b, 1990).

Root growth slows or stops soon after grasses are defoliated (Caloin et al., 1990; Jarvis & MacDuff, 1989; Crider, 1955). Root growth also is affected by the frequency and severity of defoliation. Crider (1955) reported that defoliating 50% of the grass canopy volume slowed root elongation, whereas removing 70% or more of the canopy volume caused root growth to stop. Accumulation of root mass in stargrass (*Cynodon nlemfuensis* Vanderyst var. *nlemfuensis*) was reduced by up to 97% with severe defoliation compared with an unclipped control (Alcordo et al., 1991). Mowing tallgrass prairie in the Kansas Flint Hills three or six times per year for 2 yr reduced root mass by 23 and 32%, respectively, compared with unmowed prairie (Turner et al., 1993). Carman and Briske (1982) reported that severely clipping parent shoots of newly initiated tillers of little bluestem [*Shizachyrium scoparium* (Michx.) Nash] reduced root initiation and growth of the dependent tillers. They speculated that defoliation stress may reduce forage production by interfering with the establishment of dependent tillers. Root respiration and nutrient uptake also decline quickly after defoliation (Johansson, 1993). These responses, however, may be tempered by the age of the plant and availability of resources. Thus, if root activity is slowed or stopped by repeated defoliation, water and nutrient uptake may be reduced also, further exacerbating defoliation stress (Turner et al., 1993). In contrast, Thornton and Millard (1996) reported that although root mass was reduced in cool-season perennial grasses in response to defoliation, nitrogen uptake per unit of root weight increased for some species partially offsetting the loss of root mass.

The immediate responses to defoliation are followed by the long-term process of recovering a positive carbon balance, metabolic adjustment of the remaining organs, and rebuilding the photosynthetic area (Richards, 1993). Metabolic adjustment includes an increase in photosynthetic rate of the remaining leaves (compensatory photosynthesis, defined as increased photosynthesis in leaves of defoliated plants relative to leaves of a similar age on undefoliated plants; Nowak & Caldwell, 1984; Baysdorfer & Basham, 1985). Compensatory photosynthesis may be related to a delay or halt in the normal ontogenetic decline of photosynthesis in leaves (Nowak & Caldwell, 1984; Wardlaw, 1990), exposure of shaded leaves to greater levels of light, or other resources becoming more available to the remaining leaves (Pearcy et al., 1987). Compensatory photosynthesis may be induced by increases in leaf N, carboxylase activity and amount, electron transport, or stomatal conductance (Briske & Richards, 1995).

The location and activity of remaining meristems of the defoliated plant, plant morphology, and morphological plasticity of the plant affect the long-term process of rebuilding photosynthetic area (Chapman & Lemaire, 1993). Variations in morphology among species, including meristem location and activity, are important management considerations to minimize plant stress after clipping (Briske & Richards, 1995). For example, the differences in growth habit between bunchgrass species versus stoloniferous or rhizomatous species can account for differences in rate of regrowth. More active meristems may remain below the cutting height in rhizomatous or stoloniferous species compared with bunchgrasses. Timing of tiller emergence and apical meristem elevation (synchronous or asynchronous tillering), and the ratio of reproductive to vegetative tillers produced also may affect regrowth responses. In synchronous tillering species, all tillers may develop and elevate apical meristems at once, whereas in asynchronous tillering species, this process is spread out during a longer period (Chapman & Lemaire, 1993). Improperly timed defoliation of a synchronous tillering species may remove all active meristems and reduce regrowth; however, in asynchronous tillering species fewer growing points are put at risk of defoliation (Culvenor, 1993, 1994). The morphological plasticity of some grasses may enable them to develop a more decumbent canopy structure as a defoliation avoidance mechanism (Briske & Richards, 1995).

Plants may respond differently to mechanical harvest versus grazing (Jameson, 1963). Wallace (1990) reported that grazed plants of big bluestem had higher rates of photosynthesis than mechanically clipped plants and speculated that microenvironmental differences may affect responses of grasses to clipping versus grazing. The responses of warm-season grasses to mechanical harvest or grazing also may differ because grazing may leave intact tillers on plants, whereas mechanical harvest typically removes all tillers. Matches (1966) noted that leaving some tillers intact on clipped tall fescue (*Festuca arundinacea* Schreb.) increased regrowth. Intertiller resource sharing may occur in partially defoliated plants and may aid in responses to defoliation (Welker, 1987).

Remobilization and Allocation of Carbon and Nitrogen during Regrowth

New C and N compounds are preferentially allocated to active meristems in the shoot after defoliation (Johansson, 1993; Wardlaw, 1990; Baysdorfer & Basham, 1985). The active aboveground meristems are stronger sinks than the roots and enable the plant to recover quickly from defoliation stress. This imbalance in sink strength is maintained until the amount of leaf area is large enough to meet the demands of the active sinks. The plant hormone cytokinin has been suggested as a mediator of sink strength in roots and shoots and in controlling the activity of axillary meristems perhaps through stimulating photosynthesis or phloem unloading (Murphy & Briske, 1992; Voesenek & Blom, 1996).

Briske et al. (1996) demonstrated that C_4 perennial grasses that are sensitive to clipping or grazing were less flexible in allocating C between below ground organs (roots and crown) and regrowth (new leaf blades) aboveground in response to defoliation. Big bluestem, a defoliation-sensitive species, retained more C in roots than in new leaf growth after clipping than did *Bouteloua rigidiseta* (a defoliation-tolerant species), which allocated more C to new leaf growth than to root growth. Perennial grasses that were more tolerant of defoliation were more flexible in allocating C to aboveground growth allowing them to quickly reestablish photosynthesizing leaf area.

The importance of reserve carbohydrates in regrowth of forages has been questioned (Volenec & Nelson, 1994). Research in alfalfa (*Medicago sativa* L.) has shown that the rate and amount of regrowth were not related to levels of carbohydrate in roots (Volenec, 1985; Hall et al., 1988; Boyce & Volenec, 1992; Barber et al., 1996). Although storage reserves in shoot bases, rhizomes, and roots of grasses are important to regrowth after defoliation (Dankwerts, 1993), the size of the pool of active meristems remaining after defoliation may be more important (Richards & Caldwell, 1985). Busso et al. (1990) demonstrated that regrowth of drought-stressed wheatgrasses was positively correlated with carbohydrate pools; however, regrowth was enhanced only when there were active meristems available. New evidence indicates that N reserves may be as important as C reserves during regrowth of legumes (Ta et al., 1990; Lemaire et al., 1992; Hendershot & Volenec, 1993; Ourry et al., 1994) and grasses (Thornton et al., 1994; Jarvis & MacDuff, 1989; MacDuff et al., 1989; Thornton & Millard, 1996).

HARVEST MANAGEMENT STRATEGIES

Recommendations for hay harvest management of the native tallgrasses typically state that grasses should be cut when at least 45 to 60 cm tall and before the boot stage as a compromise between forage yield and quality (Moser & Vogel, 1994). A stubble remaining after clipping of 15 to 20 cm is recommended for switchgrass, big bluestem, and indiangrass (NRCS, 1991). Regrowth is harvested when at least 45 to 60 cm tall, with enough time left for regrowth in the fall to allow plants to prepare for overwintering. Harvest management for eastern gamagrass includes leaving a 20-cm stubble height and harvesting on 1 June, 1 July, and 1 August in Kansas, or at 45-d intervals in Oklahoma (USDA-NRCS, 1991).

Brejda et al. (1996) reported that eastern gamagrass harvested at 4-wk intervals produced forage with a higher crude protein concentration than that from 6-week harvests, whereas more forage was produced with the 6-wk harvest schedule. They also observed that harvesting eastern gamagrass at 4-wk intervals during a drought year reduced plant vigor the following year.

Dwyer et al. (1963) harvested switchgrass, big bluestem, little bluestem, indiangrass, and sideoats grama [*Bouteloua curtipendula* (Michx.) Torr.] at 30- or 60-d intervals or once during the season for 6 yr in Oklahoma. Grasses were clipped at a 5- or 10-cm stubble height. The greatest forage yields were obtained when plots were cut once in July each year (Fig. 9–3). Sideoats grama, big bluestem, and indiangrass maintained 60 to 100% stands in each treatment during the 6 yr; however, switchgrass was nearly eliminated in all clipping treatments after 6 yr. Generally, root production was reduced as clipping intensity and frequency increased (Fig. 9–3).

Mullahey et al. (1990) reported that little bluestem would not tolerate close (7 cm), frequent (three harvests per season) defoliation during several years in the sandhills of Nebraska. They found that a single clipping during July was optimum in terms of forage yield, tiller weight and number, and number of crown buds. They recommended that multiple defoliations in a season should not occur during consecutive years.

Forwood and Magai (1992) harvested big bluestem at 10- or 20-cm stubble heights when the canopy reached 30, 40, or 51 cm. They found that allowing a 40-cm regrowth and cutting to a 10-cm stubble provided optimum yield and quality. Allowing plants to regrow to 40 cm provided enough time to replenish reserve compounds. They limited their recommendation to the higher rainfall area of the southern Corn Belt.

Switchgrass is considered to be very sensitive to frequent or intensive defoliation (George & Oberman, 1989) because (i) it elevates the growing point well

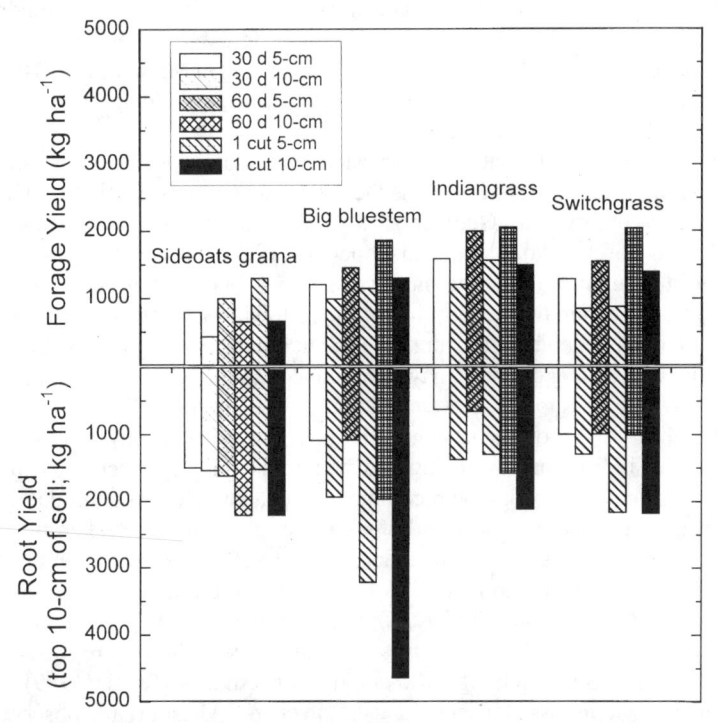

Fig. 9–3. Forage and root yields of four native warm-season perennial grasses in response to defoliation frequency and intensity in central Oklahoma. Yield data are averages of 6 yr and root data are from the sixth year. Data from Dwyer et al. (1963).

above ground during early vegetative growth (Sanderson & Wolf, 1995a), (ii) it has a high ratio of reproductive to vegetative tillers, and (iii) new growth must occur from crown buds or aerial axillary meristems (Brejda et al., 1994; Hafercamp & Copeland, 1984). Burns et al. (1984), however, reported no stand loss when 'Kanlow' switchgrass (a lowland ecotype) was grazed during late April and May for 2 yr in North Carolina. Similarly, George and Oberman (1989) recommended a mid-June defoliation of switchgrass to more evenly distribute yield and quality for grazing systems during late summer in Iowa. Anderson and Matches (1983) reported that 'Blackwell' switchgrass could be harvested at internode elongation stage with no loss of stand in central Missouri, but qualified their recommendation by limiting it to areas of higher rainfall (i.e., the humid Midwest). Van Esbroeck et al. (1994) reported that clipping frequency or intensity did not affect tiller density in three lowland cultivars of switchgrass in Texas. In that 5-yr study, tiller density remained relatively stable, whereas total annual yield declined from 10 to 6 Mg ha^{-1}.

Increased harvest frequency (one, two, three, or four harvests per season) reduced total seasonal biomass yields of switchgrass at two locations in Texas (Fig. 9–4a; Sanderson et al., 1995). The highest yields at both locations were obtained with a single harvest in September. Delaying the final fall harvest until October or November reduced yields. Similar decreases biomass yields of switchgrass were reported by Parrish and Wolf (1992) in Virginia. They suggested that switchgrass remobilized and translocated storage compounds from leaves and stems to the roots and crown during the fall, partially accounting for the yield loss. McKendrick et al. (1975) noted that the shoot weight of big bluestem and indiangrass in Kansas native prairie declined in the fall. Anderson et al. (1989) in Missouri found that total nonstructural carbohydrate concentration in above- and belowground organs of switchgrass peaked in September.

Switchgrass plants that were harvested more than once during the season in Texas had reduced first-harvest yields the following spring (Fig. 9–4b). Allowing switchgrass to regrow until October or November increased first-harvest yields in the following spring. Hafercamp and Copeland (1984) reported that intensive defoliation in autumn reduced the number of crown buds remaining and reduced switchgrass growth in the early spring of the following year. Thus, the September harvests may have depleted the number of active buds for regrowth in the spring. Alternatively, frequent clipping may have reduced root mass and, thus, limited the ability of switchgrass to obtain water and nutrients (Turner et al., 1993).

Smith (1975) reported that concentrations of reserve carbohydrates in switchgrass growing in Wisconsin were lowest when stem elongation occurred in primary spring growth and regrowth. He recommended that switchgrass should not be cut at tiller elongation because low carbohydrate reserves would weaken plants and cause stand loss. On the other hand, George et al. (1989) reported that harvest of switchgrass during stem elongation (1 June) in Iowa did not detrimentally affect plants because reserve carbohydrates could be replenished by mid summer. Neither Smith (1975) nor George et al. (1989) determined if there were active meristems remaining from which regrowth could occur. Anderson et al. (1989) clipped 'Pathfinder' switchgrass at weekly intervals in central Missouri and observed that early cutting, which left the apical meristem and some leaf blades intact, did not reduce total nonstructural carbohydrate concentrations as much as cutting at the boot

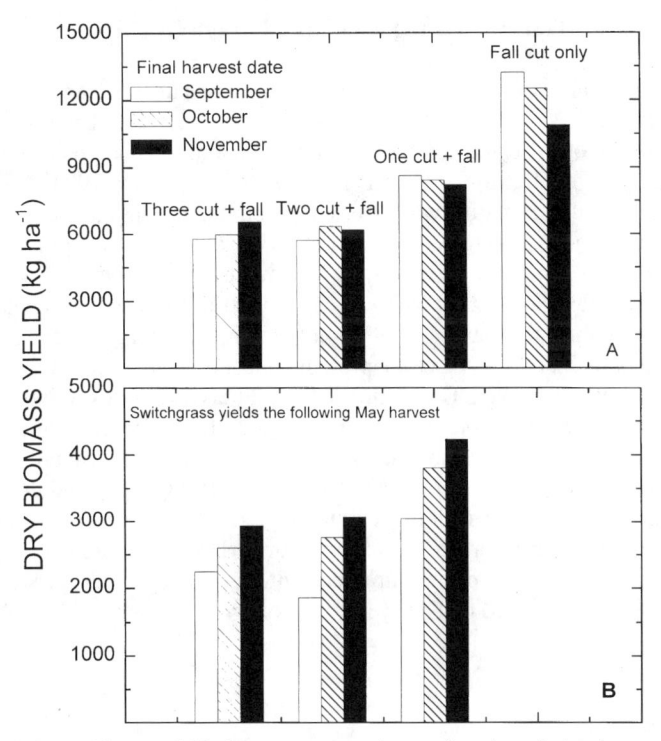

Fig. 9–4. Switchgrass biomass yield when cut one, two, three, or four times during the growing season in Texas (A). Data are averages of 3 yr and two locations (Dallas and Stephenville). Initial spring (May) forage yields of switchgrass as affected by harvest frequency the previous year (B). Data are averages of 2 yr and two locations (Sanderson, 1996, unpublished data).

stage. Late cutting removed apical meristems and left few or no intact leaf blades and forced regrowth to occur from basal (crown) buds and axillary meristems. They concluded that cutting in late summer, even as the only harvest during the year, may damage switchgrass because of few active buds available for regrowth and the limited time for plants to recover.

Harvesting for Seed and Forage

George et al. (1990) reported that harvesting switchgrass for hay or grazing in spring reduced seed yields in the fall compared with switchgrass grown for seed only. They noted that early harvests that did not removal apical meristems also reduced seed yields and speculated that the reduction in seed yield resulted from the loss of photosynthesizing leaf area and not removal of the apical meristem. Brejda et al. (1994), however, recommended cutting before stem elongation to avoid removing the apical meristem if switchgrass is to be managed for both forage and seed production. Regrowth from early harvest switchgrass came from intact apical meristems, whereas when harvest removed apical meristems, regrowth arose from crown buds and axillary meristems on the stubble. Aerial tillers arising from axillary meristems on the stubble contributed little to forage or seed yield.

Harvest Management and Forage Quality

Forage quality of perennial warm-season grasses declines as the plants mature (Anderson & Matches, 1983; Sanderson & Wolf, 1995b). The reduced quality results from both a decrease in the ratio of leaf to stem and compositional changes in stems and leaves (Buxton & Casler, 1993; Nelson & Moser, 1994).

Forage quality of native warm-season perennial grasses can be manipulated by cutting management. This often requires a compromise between forage yield, quality, and plant persistence. Harvesting less mature forage may reduce total season yield, whereas allowing forage to reach later maturity to increase yield will reduce forage quality. Brejda et al. (1996) noted that harvesting eastern gamagrass at 4-wk intervals increased the crude protein concentration because relatively less mature forage was removed. Harvesting at 6-wk intervals, however, increased forage yield but crude protein was lower. Forwood and Magai (1992) reported that harvesting big bluestem to a 10-cm stubble when regrowth was 40 cm tall resulted in optimum yield and quality. George and Obermann (1989) suggested that harvesting early growth switchgrass in Iowa would benefit forage livestock programs by providing a high quality forage in late spring or early summer, and good yields of moderate quality forage in late summer switchgrass regrowth. Anderson and Matches (1983) also noted that early harvests of switchgrass were of higher quality.

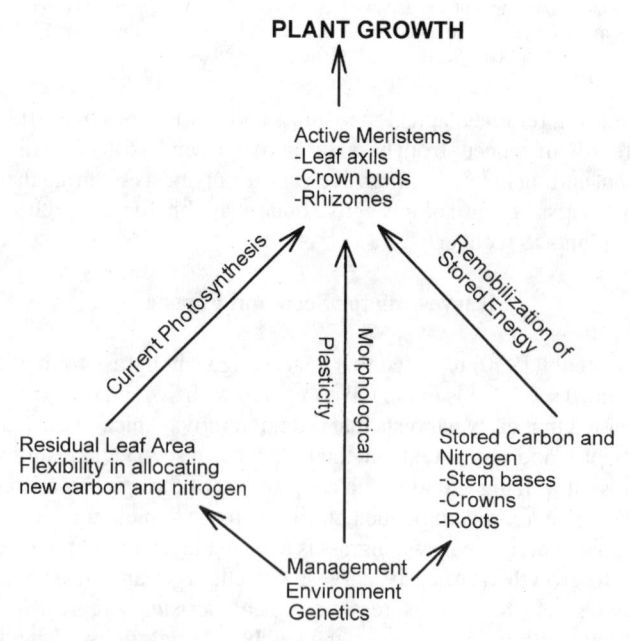

Fig. 9–5. Interrelationships among plant morphology and physiology in plant responses to defoliation (adapted from Waller et al., 1985).

SUMMARY

The native warm-season perennial grasses evolved under the evolutionary pressures of herbivory, fire, and other factors. They did not evolve in response to stresses imposed by cutting machines. Recent advances in understanding the interactions of storage compounds, plant morphology, and plant physiology help to clarify responses to defoliation. Current concepts in defoliation management recognize the importance of morphological plasticity and of maintaining a pool of active meristems that may capitalize on stored C and N compounds (Fig. 9–5). Indeed, Kemp and Culvenor (1994) stated that defoliation tolerance is best enhanced by maintaining a high density of plants and tillers that have low growing points so that regrowth is rapid after defoliation or in unfavorable environmental conditions. These traits must be incorporated into new cultivars and new management practices must be developed that increase the productivity, persistence, and forage quality of native warm-season perennial grasses.

REFERENCES

Alcordo, I.S., P. Mislevy, and J.E. Rechcigl. 1991. Effect of defoliation on root development of stargrass under greenhouse conditions. Commun. Soil Sci. Plant Anal. 22:493–504.

Anderson, B. 2000. Use of warm-season grasses by grazing livestock. p. 147–157. In K.J. Moore and B. Anderson (ed.) Native warm-season grasses: Research trends and issues. CSSA Spec. Publ. 30. CSSA and ASA, Madison, WI (this publication).

Anderson, B., and A.G. Matches. 1983. Forage yield, quality, and persistence of switchgrass and caucasian bluestem. Agron. J. 75:119–124.

Anderson, B., A.G. Matches, and C.J. Nelson. 1989. Carbohydrate reserves and tillering of switchgrass following clipping. Agron. J. 81:13–16.

Balleré, C.L., A.L. Scopel, and R.A. Sánchez. 1995. Plant photomorphogenesis in canopies, crop growth, and yield. Hortsci. 30:1172–1181.

Barber, L.D., B.C. Joern, J.J. Volenec, and S.M. Cunningham. 1996. Supplemental nitrogen effects on alfalfa regrowth and nitrogen mobilization from roots. Crop Sci. 36:1217–1223.

Baysdorfer, C., and J.A. Basham. 1985. Photosynthate supply and utilization in alfalfa. A developmental shift from a source to a sink limitation of photosynthesis. Plant Physiol. 77:313–317.

Boyce, P.J., and J.J. Volenec. 1992. Taproot carbohydrate concentrations and stress tolerance of contrasting alfalfa genotypes. Crop Sci. 32:757–761.

Brejda, J.J., J.R. Brown, T.E. Lorenz, J. Henry, J. Reid, and S.R. Lowry. 1996. Eastern gamagrass responses to different harvest intervals and nitrogen rates in northern Missouri. J. Prod. Agric. 9:130–135.

Brejda, J.J., J.R. Brown, G.W. Wyman, and W.K. Schumacher. 1994. Management of switchgrass for forage and seed production. J. Range Manage. 47:22–27.

Briske, D.D., T.W. Boutton, and Z. Wang. 1996. Contribution of flexible allocation priorities to herbivory tolerance in C_4 perennial grasses: an evaluation with ^{13}C labeling. Oecologia 105:151–159.

Briske, D.D., and J.H. Richards. 1995. Plant responses to defoliation: A physiological, morphological, and demographic evaluation. p. 635–670. In D.J. Bedunah and R.E. Sosebee (ed.) Wildland plants: Physiological ecology and developmental morphology. Soc. Range Manage., Denver, CO.

Burns, J.C., R.D. Mochrie, and D.H. Timothy. 1984. Steer performance from two perennial *Pennisetum* species, switchgrass, and a fescue-'Coastal' bermudagrass system. Agron. J. 76:795–800.

Busso, C.A., J.H. Richards, and N.J. Chatterton. 1990. Nonstructural carbohydrates and spring regrowth of two cool-season grasses: Interaction of drought and clipping. J. Range Manage. 43:336–343.

Buxton, D.R., and M.D. Casler. 1993. Environmental and genetic effects on cell wall composition and digestibility. p. 685–714. In H.G. Jung et al. (ed.) Forage cell wall structure and digestibility. ASA, CSSA, and SSSA, Madison, WI.

Caloin, M., B. Clement, and S. Herrmann. 1990. Regrowth kinetics of *Dactylis glomerata* following defoliation. Ann. Bot. 66:397–405.

Carman, J.G., and D.D. Briske. 1982. Root initiation and root and leaf elongation of dependent little bluestem tillers following defoliation. Agron. J. 74:432–435.
Chapman, D.F., and G. Lemaire. 1993. Morphogenetic and structural determinants of plant regrowth after defoliation. p. 95–104. *In* J.R. Crush et al. (ed.) Proc. 17th Int. Grassl. Congr., Palmerston North, New Zealand. 8–21 Feb., 1993. New Zealand Grassland Assoc., Palmerston, North.
Crider, F.J. 1955. Root growth stoppage resulting from defoliation of grasses. USDA Tech. Bull. 1102. U.S. Gov. Print. Office, Washington, DC.
Culvenor, R.A. 1993. Effect of cutting during reproductive development on the regrowth and regenerative capacity of the perennial grass, *Phalaris aquatica* L., in a controlled environment. Ann. Bot. 72:559–568.
Culvenor, R.A. 1994. The persistence of five cultivars of *Phalaris* after cutting during reproductive development in spring. Aust. J. Agric. Res. 45:945–962.
Dankwerts, J.E. 1993. Reserve carbon and photosynthesis: Their role in regrowth of *Themeda triandra*, a widely distributed subtropical graminaceous species. Func. Ecol. 7:634–641.
Dwyer, D.D., W.C. Elder, and G. Singh. 1963. Effects of height and frequency of clipping on pure stands of range grasses in north central Oklahoma. Bull. B-614. Oklahoma State Univ., Stillwater.
Forwood, J.R., and M.M. Magai. 1992. Clipping frequency and intensity effects on big bluestem yield, quality, and persistence. J. Range Manage. 45:554–559.
Frank, A.B., and A. Bauer. 1995. Phyllochron differences in wheat, barley, and forage grasses. Crop Sci. 35:19–23.
Frank, A.B., and L. Hofmann. 1994. Light quality and stem numbers in cool-season forage grasses. Crop Sci. 34:468–473.
George, J.R., and D. Oberman. 1989. Spring defoliation to improve summer supply and quality of switchgrass. Agron. J. 81:47–52.
George, J.R., D.J. Oberman, and D.D. Wolf. 1989. Seasonal trends for nonstructural carbohydrates in stem bases of defoliated switchgrass. Crop Sci. 29:1282–1287.
George, J.R., G.S. Reigh, R.E. Mullen, and J.J. Hunzak. 1990. Switchgrass herbage and seed yield and quality with partial spring defoliation. Crop Sci. 30:845–849.
Gerrish, J.R., J.R. Forwood, and C.J. Nelson. 1987. Phenological development of eleven warm season grass cultivars. p. 249–252. *In* Proc. Am. Forage and Grassl. Conf., Springfield, IL. 2–5 Mar. 1987. Am. Forage and Grassl. Council, Georgetown, TX.
Gold, W.G., and M.M. Caldwell. 1989a. The effects of spatial patterns of defoliation on regrowth of a tussock grass: I. Growth responses. Oecologia 80:289–296.
Gold, W.G., and M.M. Caldwell. 1989b. The effects of spatial patterns of defoliation on regrowth of a tussock grass: II. Canopy gas exchange. Oecologia 81:437–442.
Gold, W.G., and M.M. Caldwell. 1990. The effects of spatial patterns of defoliation on regrowth of a tussock grass: III. Photosynthesis, canopy structure, and light interception. Oecologia 82:12–17.
Hafercamp, M.R., and T.D. Copeland. 1984. Shoot growth and development of Alamo switchgrass as influenced by mowing and fertilization. J. Range Manage. 37:406–412.
Hall, M.H., C.C. Sheaffer, and G.H. Heichel. 1988. Partitioning and mobilization of photoassimilate in alfalfa subjected to water deficits. Crop Sci. 28:964–969.
Hendershot, K.L., and J.J. Volenec. 1993. Taproot nitrogen accumulation and use in overwintering alfalfa (*Medicago sativa* L.). J. Plant Physiol. 141:68–74.
Holmes, M.G., and H. Smith. 1977. The function of phytochrome in the natural environment: II. The influence of vegetation canopies on the spectral energy distribution of natural daylight. Photochem. Photobiol. 25:539–545.
Hyder, D.N. 1974. Morphogenesis and management of perennial grasses in the United States. p. 89–98. *In* Plant morphogenesis as the basis for scientific management of range resources. Proc. of a Workshop of the USA–Australia Rangelands Panel, Berkeley, CA. 29 Mar.–5 Apr. 1971. USDA Misc. Publ. 1271. U.S. Gov. Print. Office, Washington, DC.
Jameson, D.A., 1963. Responses of individual plants to harvesting. Bot. Rev. 29:532–594.
Jarvis, S.C., and J.H. MacDuff. 1989. Nitrate nutrition of grasses from steady-state supplies in flowing solution culture following nitrate deprivation and/or defoliation: I. Recovery of uptake and growth and their interactions. J. Exp. Bot. 40:965–975.
Johansson, G. 1993. Carbon distribution in grass (*Festuca pratensis* L.) during regrowth after cutting—utilization of stored and newly assimilated carbon. Plant Soil 151:11–20.
Kemp, D.A., and R.A. Culvenor. 1994. Improving the grazing and drought tolerance of temperate perennial grasses. NZ J. Agric. Res. 37:365–378.
Lemaire, G., M. Khaity, B. Onillon, J.M. Allirand, M. Chartier, and G. Gosse. 1992. Dynamics of accumulation and partitioning of N in leaves, stems, and roots of lucerne (*Medicago sativa* L.) in a dense canopy. Ann. Bot. 70:429–435.

MacDuff, J.H., S.C. Jarvis, and A. Mosquera. 1989. Nitrate nutrition of grasses from steady-state supplies in flowing solution culture following nitrate deprivation and/or defoliation: II. Assimilation of NO_3^- and short-term effects on NO_3^- uptake. J. Exp. Bot. 40:977–984.

Matches, A.G. 1966. Influence of intact tillers and height of stubble on growth responses of tall fescue (*Festuca arundinacea* Schreb.) Crop Sci. 6:484–487.

McLaughlin, S.B. 1993. New switchgrass biofuels research program for the southeast. p. 111–115. *In* Proc. Annual Automotive Tech. Dev. Cont. Coordination Meeting, Dearborn, MI. 2–5 Nov. 1992. Soc. Auto. Eng., Warrendale, PA.

McKendrick, J.D., C.E. Owensby, and R.M. Hyde. 1975. Big bluestem and indiangrass vegetative reproduction and annual reserve carbohydrate and nitrogen cycles. Agro-Ecosyst. 2:75–93.

Mitchell, R., and L.E. Moser. 2000. Developmental morphology and tiller dynamics of warm-season grass swards. p. 49–66. *In* K.J. Moore and B. Anderson (ed.) Native warm-season grasses: Research trends and issues. CSSA Special Publ. 30. CSSA and ASA, Madison, WI (this publication).

Moore, K.J., and L.E. Moser. 1995. Quantifying developmental morphology of perennial grasses. Crop Sci. 35:37–43.

Moser, L.E., and K.P. Vogel. 1994. Switchgrass, big bluestem, and indiangrass. p. 409–420. *In* Forages: An introduction to grassland agriculture. Vol. 1. Iowa State Univ. Press, Ames.

Mullahey, J.J., S.S. Waller, and L.E. Moser. 1990. Defoliation effects on production and morphological development of little bluestem. J. Range Manage. 43:497–500.

Murphy, J.S., and D.D. Briske. 1992. Regulation of tillering by apical dominance: Chronology, interpretive value, and current perspectives. J. Range Manage. 45:419–429.

Nelson, C.J. 1994. Photosynthesis and carbon metabolism. p. 31–43. *In* Forages: An introduction to grassland agriculture. Vol. 1. Iowa State Univ. Press, Ames.

Nelson, C.J., and L.E. Moser. 1994. Plant factors affecting forage quality. p. 115–154. *In* G.C. Fahey, Jr. (ed.) Forage quality, evaluation, and utilization. ASA, CSSA, and SSSA, Madison, WI.

Nowak, R.S., and M.M. Caldwell. 1984. A test of compensatory photosynthesis in the field: Implications for herbivory tolerance. Oecologia 61:311–318.

Ourry, A., T.H. Kim, and J. Boucaud. 1994. Nitrogen reserve mobilization during regrowth of *Medicago sativa* L. Relationships between availability and regrowth yield. Plant Physiol. 105:831–837.

Parrish, D.J., and D.D. Wolf. 1992. Managing switchgrass for sustainable biomass production. p. 34–39. *In* J. S. Cundiff (ed.) Proc. of and Alternative Energy Conf. Liquid Fuels from Renewable Resources, Nashville, TN. 14–15 Dec. 1992. Am. Soc. Agric. Eng., St. Joseph, MI.

Pearcy, R.W., O. Bjorkman, M.M. Caldwell, J.E. Keeley, R.K. Monson, and B.R. Strain. 1987. Carbon gain by plants in natural environments. Bioscience 37:21–29.

Richards, J.H. 1993. Physiology of plants recovering from defoliation. p. 85–94. *In* J.R. Crush et al., (ed.) Proc. 17th Int. Grassl. Congr., Palmerston North, New Zealand, and Rockhapton, Queensland, Australia. 8–21 Feb., 1993. New Zealand Grassland Assoc., Palmerston North.

Richards, J.H., and M.M. Caldwell. 1985. Soluble carbohydrates, concurrent photosynthesis and efficiency in regrowth following defoliation: A field study with *Agropyron* species. J. Appl. Ecol. 22:907–920.

Sanderson, M.A. 1992. Morphological development of switchgrass and kleingrass. Crop Sci. 84:415–419.

Sanderson, M.A., M.A. Hussey, W.R. Ocumpaugh, C.R. Tischler, J.C. Read, and R.L. Reed. 1995. Evaluation of switchgrass as a sustainable bioenergy crop in Texas. p. 253–260. *In* Proc. Second Biomass Conf. of the Americas, Portland, OR. 21–24 Aug. 1995. Natl. Renewable Energy Lab., Golden, CO.

Sanderson, M.A., R.L. Reed, S.B. McLaughlin, S.D. Wullschleger, B.V. Conger, D.J. Parrish, D.D. Wolf, C. Taliaferro, A.A. Hopkins, W.R. Ocumpaugh, M.A. Hussey, J.C. Read, and C.R. Tischler. 1996. Switchgrass as a sustainable bioenergy crop. Biores. Tech. 56:83–93.

Sanderson, M.A., and D.D. Wolf. 1995a. Morphological development of switchgrass in diverse environments. Agron. J. 87:908–915.

Sanderson, M.A., and D.D. Wolf. 1995b. Switchgrass biomass composition during morphological development in diverse environments. Crop Sci. 35:432–438.

Smith, D. 1975. Trends of nonstructural carbohydrates in the stem bases of switchgrass. J. Range Manage. 28:389–391.

Ta, T.C., F.D.H. MacDowall, and M.A. Faris. 1990. Utilization of carbon and nitrogen reserves of alfalfa roots in supporting N_2-fixation and shoot regrowth. Plant Soil 127:231–236.

Thornton, B., and P. Millard. 1996. Effects of severity of defoliation on root functioning in grasses. J. Range Manage. 49:443–447.

Thornton, B., P. Millard, and E.I. Duff. 1994 Effects of nitrogen supply on the source of nitrogen used for regrowth of laminae after defoliation of four grasses. New Phytol. 128:615–620.

Turner, C.L., T.R. Seastedt, and M.I. Dyer. 1993. Maximization of aboveground grassland production: The role of defoliation frequency, intensity, and history. Ecol. Appl. 3:175–186.

USDA-NRCS. 1991. Native perennial warm-season grasses for forage in southeastern United States (except Florida). Switchgrass, indiangrass, big bluestem, little bluestem, sideoats grama, maidencane, coastal panicgrass, and eastern gamagrass. Ecological sciences and planning staff, Ft. Worth, TX, February, 1991. USDA-NRCS, Washington, DC.

Van Esbroeck, G., E.C. Holt, M.A. Hussey, and M.A. Sanderson. 1994. Effects of clipping on switchgrass persistence and productivity. p. 166. *In* Agronomy abstract. ASA, Madison, WI.

Voesenek, L.A. C.J., and C.W.P.M. Blom. 1996. Plants and hormones: An ecophysiological view on timing and plasticity. J. Ecol. 84:111–119.

Vogel, K.P. 1996. Energy production from forages (or American agriculture–back to the future). J. Soil Water Conserv. 51:137–139.

Volenec, J.J. 1985. Leaf area expansion and shoot elongation of diverse alfalfa germplasms. Crop Sci. 25:822–827.

Volenec, J.J., and C.J. Nelson. 1994. Forage crop management: Application of emerging technologies. p. 3–29. *In* Forages: The science of grassland agriculture. Vol. 3. Iowa State Univ. Press, Ames.

Wallace, L.L. 1990. Comparative photosynthetic responses of big bluestem to clipping versus grazing. J. Range Manage. 43:58–61.

Waller, S.S., L.E. Moser, P.E. Reece, and G.A. Gates. 1985. Understanding grass growth: The key to profitable livestock production. Trabon Printing Co., Kansas City, MO.

Wardlaw, I.F. 1990. The control of carbon partitioning in plants. New Phytol. 116:341–381.

Weaver, J.E., and V.H. Hougen. 1939. Effect of frequent clipping on plant production in prairie and pasture. Am. Midland Nat. 21:396–414.

Welker, J.M., D.D. Briske, and R.W. Weaver. 1987. Nitrogen-15 partitioning within a three generation tiller sequence of the bunchgrass *Schizachyrium scoparium*: Response to selective defolation. Oecologia 24:330–334.

10 Use of Warm-Season Grasses by Grazing Livestock

Bruce E. Anderson

University of Nebraska
Lincoln, Nebraska

Warm-season grasses like big bluestem (*Andropogon gerardii* Vitman), switchgrass (*Panicum virgatum* L.), and indiangrass [*Sorghastrum nutans* (L.) Nash] have great potential as summer pasture grasses throughout much of the Corn Belt and Great Plains. While cool-season grasses, such as tall fescue (*Festuca arundinacea* Schreb.), smooth brome (*Bromus inermis* Leyss.), and orchardgrass (*Dactylis glomerata* L.), often are nearly dormant and unproductive during the heat of midsummer, warm-season grasses thrive under these conditions.

Warm-season grasses start growth about 4 to 6 wk later in spring than do cool-season grasses. As a result, spring soil moisture is conserved. From 60 to 90% of the annual growth of warm-season grasses usually occurs during June through August. In contrast, >60% of the growth of cool-season grasses often occurs before June (Rountree et al., 1974).

Like cool-season grasses, warm-season grasses grow best on well-drained, fertile soils and they respond well to proper fertilization, weed control, and defoliation management. Warm-season grasses thrive at 30 to 35°C, but grow very slowly below 20°C. They use less water than cool-season grasses to produce similar growth (Downes, 1969). They also are more efficient in N use (Brown, 1978) and grow better than cool-season grasses on P-deficient soils (Morris et al., 1982). Thus, warm-season grasses often are grown on soils with growth limitations because they perform better than cool-season grasses on those sites.

Switchgrass tolerates poorly drained soils, flooding, perched water tables, and salinity better than many other commonly seeded warm-season grasses (Duke, 1978). Big bluestem is more drought tolerant than other tall warm-season grasses and thus may be better adapted to excessively drained soils with low water-holding capacity. Even better adapted to droughty sites are two mid-grasses, little bluestem [*Schizachyrium scoparium* (Michx.) Nash] and sideoats grama [*Bouteloua curtipendula* (Michx.) Torr.], which often are seeded in mixtures with the tall warm-season grasses.

Native warm-season grasses have great potential to provide desirable summer grazing when cool-season grasses are less productive; however, warm-season grasses differ from cool-season grasses in their response to grazing. Warm-season

---

Copyright © 2000. Crop Science Society of America and American Society of Agronomy, 677 S. Segoe Rd., Madison, WI 53711, USA. *Native Warm-Season Grasses: Research Trends and Issues.* CSSA Special Publication no. 30.

grasses evolved under grazing, but it was intermittent so they are not well adapted to lengthy periods of continuous stocking or close, frequent defoliation. And their nutritional characteristics provide unique challenges and opportunities for grazers. Thus, research about grazing these grasses has focused on solving these problems and identifying useful roles for warm-season grasses in livestock production systems.

GRAZING RESEARCH

Systems

Grazing research with warm-season grasses initially focused on summer gain potential and/or opportunities to use warm-season grasses in grazing systems. Early Nebraska studies showed that average daily gains (ADG) from late April through early November increased 38% when steers (*Bos taurus*) grazed cool-season grasses in spring and autumn and warm-season grasses during summer compared with grazing cool-season grasses season-long (Conard & Clanton, 1963). Improved season-long gains were due to a three-fold increase in daily gain during summer for steers grazing warm-season grasses. Subsequent studies averaged 31% more gain animal^{-1} and 96% more gain ha^{-1} when warm-season grasses were grazed by yearling steers during summer than stockpiled smooth brome (Klopfenstein & Lewis, 1988). Cows (0.44 vs 0.15 kg) and calves (0.93 vs 0.78 kg) also had higher ADG when they sequentially grazed smooth brome, switchgrass, big bluestem, and smooth brome again than when they grazed smooth brome continuously (Ward, 1988).

In Missouri, switchgrass and caucasian bluestem [*Bothriochloa caucasia* (Trin.) C.E. Hubb.] produced daily and per hectare gains during summer similar to tall fescue plus fescue hay without the need to harvest hay (Matches et al., 1975; Matches et al.,1982; Anderson, 1986). Summer gains often exceeded 0.5 kg d^{-1} in a region characterized by summer weight loss by cattle when endophyte-infected tall fescue was grazed alone. Similarly, Oklahoma studies with switchgrass and sand bluestem [*Andropogon gerardii* var. *paucipilus* (Nash) Fern.] produced ADG from 0.37 to 0.62 kg (Dwyer & Elder, 1964).

Beef yearling steers gained an average of 0.9 kg daily and ranged from 0.7 to 1.08 kg hd^{-1} d^{-1} from switchgrass, big bluestem, indiangrass, and sideoats grama in South Dakota during 3 yr (Krueger & Curtis, 1979). Gain per hectare ranged from 112 to 147 kg. In western Iowa (Wedin & Fruehling, 1977), both switchgrass and aftermath smooth brome produced slightly >0.6 kg ADG during mid-summer but switchgrass provided more steer days of grazing (139 vs. 96 d) and higher gains per hectare (225 vs. 148 kg). Subsequent studies in Iowa demonstrated that higher stocking rates could be maintained using a grazing system of smooth brome and switchgrass grazed alternately compared with season-long grazing of smooth brome. Within this system, switchgrass produced higher ADG than smooth brome (0.75 vs. 0.67 kg) but lower carrying capacity (452 vs. 507 steer d ha^{-1})(Wedin et al., 1980).

Similar results grazing warm-season grasses alone or in grazing systems have been recorded in North Carolina (Burns et al., 1984), Pennsylvania (Fairbairn et al., 1985), West Virginia (Reid & Jung, 1985), and Illinois (Kaiser et al., 1986). In contrast, grazing studies in Wisconsin showed no advantage to sequentially grazing Kentucky bluegrass (*Poa pratensis* L.) with switchgrass compared with grazing bluegrass season-long (Smart et al., 1995). Under Wisconsin conditions, there was no slump in pasture and animal productivity during summer from bluegrass.

Management

Until recently, most grazing management recommendations for warm-season grasses in the Corn Belt and Great Plains were similar to those used with cool-season grasses, although there were some attempts to use deferred rotation grazing schemes similar to those employed in native rangelands (Smith & Owensby, 1978). Current grazing research has shown, however, that in these regions, warm-season grasses respond differently than cool-season grasses to grazing strategies such as grazing frequency, intensity, duration, and timing.

In Missouri, big bluestem was stocked continuously or rotationally, with rotational stocking designed to provide either a 7-d grazing period with a 14-d rest period or 1-d grazing period with a 20-d rest period (Gerrish et al., 1994). Although rotational stocking provided more forage mass than continuous stocking all 3 yr of the study, virtually all big bluestem plants died in all pastures and were replaced by volunteer summer annual grasses. Inadequate rest period was suspected to be part of the problem. Stand decline was most rapid with continuous stocking and following trampling damage during wet weather. Intake by lactating beef cows grazing these pastures and in vitro dry matter digestibility (IVDMD) of the forage consumed was greater when cows were rotated daily compared with continuous stocking; both intake and IVDMD declined from Day 1 to Day 7 during the 7-d grazing periods (Morrow et al., 1994).

Switchgrass was grazed at three stocking rates in Alabama using continuous stocking (Maposse et al., 1995) or an eight-paddock rotation (Maposse et al., 1996). As stocking rate increased, ADG decreased in both single-year studies, primarily late in the grazing season; however, there were no differences between grazing methods.

Iowa studies have shown large advantages to intensive early stocking of switchgrass and big bluestem compared with waiting to start continuous stocking late in the grazing season (George et al., 1996). Continuous stocking began in late June or mid-July. Intensive early grazing occurred for 2 wk or less in late May or mid-June and was followed by a 4-wk rest period and then a second grazing period. During 3 yr, steer ADG was 1.10 vs. 0.85 kg for switchgrass and 1.30 vs. 1.10 kg for big bluestem from rotational vs. continuous stocking, respectively, while gain per hectare was 529 vs. 192 kg for switchgrass and 396 vs. 220 kg for big bluestem. Improved gains were attributed to less waste from trampling with intensive early stocking as well as less mature plants and greater leaf-to-stem ratios for grazing, resulting in higher forage quality at time of grazing.

Nebraska studies recently have compared continuous to rotational stocking at three stocking rates during a 6-yr grazing trial using a five-species mixture of warm-season grasses (Anderson, 1996a, 1999). Average daily gains tended to decline as stocking rate increased, but the decline was steepest with continuous stocking. During the first 2 yr, gains per hectare were greatest with continuous stocking, but rotational stocking produced progressively higher gains per hectare in relation to continuous stocking in later years, especially as stocking rate increased. Rotational stocking extended the grazing season slightly and gains tended to be influenced less by stocking rate.

Forage Use and Quality

Warm-season (C_4) grasses generally are found to be lower in forage quality, i.e., fiber digestion, in vitro dry matter digestibility (IVDMD), or crude protein concentration, than cool-season (C_3) grasses (Moore & Buxton, 1999; Wilson et al., 1983; Brown, 1978). Differences in leaf anatomy associated with the C_3 and C_4 photosynthetic pathways may be partly responsible (Laetsch, 1974). Leaves of C_4 grasses contain more bundle sheath cells and fewer mesophyll cells than C_3 leaves (Morgan & Brown, 1979). Since mesophyll cell walls degrade more rapidly than bundle sheath cells (Akin et al., 1973), these anatomical differences may explain why warm-season grasses often are lower in IVDMD than cool-season grasses.

But, performance of animals consuming these grasses often is greater than might be expected based on standard laboratory analyses of forage quality (Abrams et al, 1981; Reid et al., 1988). When big bluestem and switchgrass were fed to sheep, IVDMD underestimated in vivo digestible dry matter (DDM) by approximately 17 percentage units (Griffin et al., 1980). When similar hays were fed to cattle and sheep (*Ovis aries*), cattle tended to digest the warm-season grasses more completely than sheep, especially at higher levels of digestibility (Vona et al., 1984). Cattle also tended to consume more forage (body weight$^{0.75}$) than did sheep when digestibility was relatively high, but there was little difference in intake of low digestibility forage.

Akin and Burdick (1975) demonstrated that C_4 grass tissue degrades in vitro more slowly than C_3 grass tissue due to differences in tissue types. Lower IVDMD should result in lower animal gains; however, intake of warm-season grasses by cattle often is higher than that of cool-season grasses at comparable digestibility levels (Minson, 1981), even though the neutral detergent fiber (NDF) levels of warm-season grasses also are higher (Vona et al., 1984). In fact, mean daily intake of NDF by cattle consuming warm-season grasses was 38 and 15% higher (body weight$^{0.75}$) than when consuming legumes or cool-season grasses, respectively (Reid et al., 1988).

It is unlikely that cell walls of warm-season grasses degrade more rapidly in the rumen or have shorter rumen retention times than those of cool-season grasses. A more likely explanation of unexpectedly high intakes is an ability of cattle to accommodate greater rumen fill with warm-season grasses (Reid et al., 1988). Vaughn (1997) measured rumen fill of cattle grazing smooth brome, switchgrass, or big bluestem using total rumen evacuation. He found rumen fill (dry matter basis) with smooth brome was 1.1% of body weight compared with 1.5% for the warm-season grasses.

Protein characteristics of warm-season grasses also differ from cool-season grasses. For a more detailed description of these differences, see Redfearn and

Hollingsworth-Jenkins (1999). Leaves of warm-season grasses contain less crude protein (CP) than cool-season grasses (Brown, 1978). Levels of undegraded intake protein (UIP), a.k.a. bypass protein, measured on a CP basis tend to be higher in warm-season grasses and levels of degradable intake protein (DIP), a.k.a. rumen degradable protein, tend to be lower. For example, Mullahey et al. (1992) determined that UIP of whole-plant switchgrass was greater than smooth brome (31.8 vs. 22.3 g kg^{-1} CP) when calculated on a CP basis and declined with maturity for both grasses. Redfearn et al. (1995) estimated UIP of switchgrass and big bluestem leaves to be 3.4- to 6.2-fold higher than smooth brome (CP basis). But when calculated on a dry matter basis, Kirch (1995) found UIP was similar for the same three grasses. Both Kirch (1995) and Mullahey et al. (1992) found that UIP (DM basis) declined in warm-season grasses as they matured but its concentration was relatively stable in smooth brome, a cool-season grass.

Cool-season grasses appear to contain sufficient DIP to maintain proper rumen fermentation but are deficient in UIP for maximum animal performance. Anderson et al. (1988) increased daily gains of yearling steers grazing smooth brome by supplementing UIP, even though whole-plant samples contained an average of 10.4% CP during spring and 13.4% CP during fall. Esophageally collected samples from similar pastures in the same year contained an average of 15.6% CP. Similar supplementation of beef cows grazing smooth brome pastures in eastern Nebraska increased cow milk production and calf gains (Blasi et al., 1991).

In contrast, when warm-season grasses are digested, relatively large amounts of soluble protein escape rumen degradation and low amounts of rumen ammonia are produced (Hunter & Siebert, 1985). Thus, diets based on warm-season grasses occasionally may be deficient in DIP. Warm-season grass diets likely have adequate UIP while plants remain relatively immature but often become deficient as plants mature. When beef yearlings grazing warm-season grasses experienced low levels of rumen ammonia (<5 mg dL^{-1}), Hafley et al. (1993) increased ADG by 0.09 kg by supplementing them with DIP. Supplementation with UIP alone provided no improvement but a combination of UIP and DIP increased gains 0.13 kg compared with no supplements. Supplements providing a combination of UIP and DIP also increased gains of cattle grazing bermudagrass [*Cynodon dactylon* (L.) Pers.] in Texas (Grigsby et al., 1988). Blasi et al. (1991) found no benefit to supplementing lactating beef cows with UIP alone while grazing big bluestem.

At least part of the reason a relatively high portion of the protein in warm-season grasses escapes rumen degradation is that much of the protein is located inside slowly degraded bundle sheath cells. Ribulose-1,5-bisphosphate carboxylase (RuBPCase) is an abundant, highly soluble protein found only in bundle sheath cells of warm-season grasses (Ku et al., 1979). Miller et al. (1996) found RuBPCase in omasal and fecal material from cattle grazing switchgrass and big bluestem, indicating that some of this protein avoids rumen degradation. They suggested this was due to protection within the bundle sheath cells and may be available as UIP in the lower tract. Vaughn (1997), using an enzyme-linked immunosorbent assay, conducted trials with grazing cattle and estimated that 10 to 40% of the RuBPCase in ingested big bluestem and switchgrass escaped rumen degradation and disappeared in the lower tract. Stage of growth affected results inconsistently; as plants matured, the amount of protein that disappeared due to rumen degradation declined while protein disappearance in the lower tract and protein that escaped the entire tract increased. Few differences occurred during a second year.

Fertilization of warm-season grass pastures with N may improve rumen fermentation. Chapman and Kretschmer (1964) used N fertilization to increase CP concentration and intake of low protein C_4 forages by steers. In switchgrass, big bluestem, and indiangrass, 65% of the additional CP associated with N fertilization was DIP (Cuomo & Anderson, 1996).

SUGGESTED GRAZING METHODS

Proper grazing management is crucial to maintain dense, productive, nutritious stands of warm-season grasses. While plant response to grazing is similar, warm-season grasses should be grazed differently than cool-season grasses. Major differences include length of rest, sensitivity to severe grazing (especially late season), and grazing before elevation of apical meristems (Anderson, 1996b; Mitchell et al., 1994). The following suggested grazing methods appear appropriate throughout the Great Plains and Corn Belt.

Tall and medium height warm-season grasses need lengthy rest periods (40+ days) via rotational grazing when livestock reduce stubble to <20 cm and consume nearly all the leaves. If stubble height can be maintained between 25 and 40 cm, lengthy grazing periods or continuous stocking can be effective. Stubble should not be grazed shorter than 20 cm after early September.

Stem management is nearly as important as leaf management with tall warm-season grasses. Stems become unpalatable, have low digestibility and protein, and reduce intake. Removing the shoot apices by grazing during jointing and stem elongation will reduce mature stem development while stimulating leaf growth and consumption.

Stems elongate and plants mature at different times for different warm-season grasses (Branson, 1953). Switchgrass develops stems 2 to 3 wk earlier than many warm-season grasses; indiangrass is one of the latest to develop stems. When several warm-season grasses are mixed together in the same pasture, switchgrass becomes stemmy and less palatable earlier in summer than other warm-season grasses. Livestock refuse to eat switchgrass and selectively graze less stemmy grasses, especially preferring big bluestem. Palatable grasses will be overgrazed while switchgrass is underused in the same pasture.

Because switchgrass has such unique affects on grazing, management will be discussed using three broad categories: pure switchgrass, mixed stands with switchgrass, and mixed stands without switchgrass.

Pure Switchgrass

Switchgrass Must be Grazed before Seedstalks Develop

Before seedstalks develop, forage quality is high and palatability good. After seedheads emerge, nutrient levels become low and switchgrass becomes unpalatable as pasture. Animals are reluctant to eat mature switchgrass and may refuse it entirely if other feed is available (Mitchell et al., 1994).

Grazing of switchgrass must begin when it becomes ready to graze, regardless of how much grazing potential remains on cool-season pastures. It is better to graze switchgrass when it is ready and then graze the remaining cool-season grass later in the summer than to finish grazing the cool-season grass first and let switchgrass become stemmy. If switchgrass becomes stemmy before grazing begins, it should be cut for hay and the regrowth grazed about 45 d later.

Several options are available for grazing switchgrass. One option is to begin grazing when switchgrass is 20 to 25 cm tall by stocking the pasture so livestock will consume switchgrass at the same rate that it grows. Livestock will graze off the tops of switchgrass plants rather uniformly if coarse stems have not started to form. Plant height should be kept between 20 and 40 cm for 6 to 8 wk, then remove livestock for 30 to 45 d. Stocking switchgrass too heavily and moving animals to other pastures sooner than planned is better than stocking too lightly and having abundant seedheads develop. Regrowth may be grazed (it may be fairly stemmy with good growing conditions) to a stubble height no shorter than 20 cm.

It is difficult to predict switchgrass growth rate and to stock it to maintain 20 to 40 cm of stubble. Thus, an easier grazing method is to begin grazing when switchgrass is 25 to 30 cm tall and stock heavily for 2 to 3 wk until there is about 10 to 15 cm of stubble. Then livestock should be removed and switchgrass allowed to recover for 40 or more days. If at least 30 cm of regrowth occurs, it may be grazed again but to no less than 20-cm stubble height. Grazing that removes young stems in early summer will reduce and delay switchgrass heading and provide higher quality regrowth forage later in the grazing season.

Switchgrass is difficult to use as the only forage source for all of June, July, and August because of its rapid growth rate and early development of seedstalks. It is much easier to use switchgrass during just two of these three summer months. Either graze switchgrass uniformly in June and July or graze completely in June and graze regrowth in August. Avoid stemmy, mature growth.

Mixed Stands with Switchgrass

The biggest challenge in these stands is reducing selective grazing so switchgrass will be used and more palatable grasses will not be overgrazed (Anderson, 1996b). As switchgrass begins to form seedstalks that are not grazed, use the palatable grasses to judge stubble height.

Selective grazing can be controlled by management. If only one warm-season grass pasture is available, grazing options are similar to pure switchgrass. When grass gets 20 to 25 cm tall, stock so livestock consume growth about as fast as it grows. Livestock will graze fairly evenly for about 3 or 4 wk until switchgrass gets less palatable. Grazing usually can continue another 2 or 3 wk with yearlings gaining 0.7 to 1.2 kg d^{-1} during this total period.

At this point, if 20 or more cm of growth remain on the more palatable grasses, grazing can continue until stubble height is only 10 to 15 cm; then end grazing for the year. Alternatively, the warm-season grass pasture can be rested for at least 4 wk, followed by light grazing to no shorter than a 20-cm stubble. If stubble is shorter than 20 cm at this point, rest the pasture for 4 wk *plus* one more week for every 2.5

cm of stubble below 20 cm. Then regrowth can be grazed lightly until the end of the growing season to no shorter than a 20-cm stubble.

The other option for one mixed warm-season grass pasture is to stock the pasture heavily enough when grass gets 30 to 40 cm tall to reduce growth to 15 cm of stubble within 3 wk. This grazing must begin early and use switchgrass before stems develop. Growth remaining on cool-season pasture should be left if necessary to accomplish this early grazing. Then return to cool-season pastures for about 6 wk to use their remaining growth and permit warm-season grasses to regrow. The regrowth of the warm-season grasses can be grazed until either cool-season grasses are ready for fall use or until warm-season grass stubble is 20 cm.

A better way to use mixed stands is to rotationally graze. Alternate which paddock is grazed first each year. Grazing should begin when grass is 20 to 25 cm tall. Each paddock should be grazed only a few days so that livestock begin grazing the final paddock as switchgrass is jointing, *before* the boot stage. Graze this last paddock until 15 cm of stubble remains before continuing the rotation. Then graze each of the rest of the paddocks until about 15 cm of stubble remains on the palatable grasses to reduce seedstalks in regrowth. All paddocks grazed to a 15-cm stubble height should be rested for 40+ days before regrazing, leaving at least 20 cm of stubble going into winter.

Mixed Stands Without Switchgrass

Because most other warm-season grasses form seedstalks later than switchgrass, they may be ready to graze at a more convenient time than switchgrass, relative to growth of cool-season pastures. Also, more flexibility is available when switchgrass is not in the stand. Still, proper management remains necessary to maintain grass stands, high forage quality and yield, and good animal performance.

Mixed stands can be stocked continuously all summer if livestock begin grazing when grass is 25 to 30 cm tall, grazing distribution is uniform (unlikely in large pastures), and stocking rate or adjustments in animal numbers maintain a consistent amount of leaf area and stubble. Be sure at least 15 to 20 cm of stubble remains for winter.

Rotationally stocking mixed stands of warm-season grasses provides more uniform grazing distribution and it is not as sensitive to stocking rate as continuous stocking. Begin grazing when grass is 20 to 30 cm tall.

One method of rotational stocking involves grazing each paddock no longer than 2 to 3 wk to a stubble height about equal to one-half the original plant height. This method works well for a system with two to four paddocks. Another method that works well with just a few paddocks is heavily grazing each paddock to 10 to 15 cm. Then rest each paddock for at least 40 d, often longer. Delay further grazing of any paddock until regrowth is at least 30 cm tall. Remove livestock during subsequent grazing periods so the average stubble height is at least 15 cm. Conclude grazing so at least 15 to 20 cm of stubble will be on each paddock going into winter.

With six or more paddocks, graze each paddock lightly at first, removing no more than 25% of the growth in 1 to 3 d. Complete the first rotation within 2 or 3 wk. The next grazing can be done the same way, but as seedstalks begin to develop and plants are noticeably jointing, graze to a shorter 15-cm stubble. Adjust rest periods so paddocks grazed short receive at least 40 d of recovery; paddocks with 25+ cm of stubble may need only 25 to 30 d before regrazing.

CONCLUSIONS

Warm-season grasses provide abundant, high-quality pasture during summer in the Great Plains and Corn Belt regions. Although their basic physiological and morphological traits are similar to cool-season grasses, differences in plant anatomy, protein degradation, and growth characteristics cause responses to grazing to differ. Ingested warm-season grasses commonly release inadequate amounts of N into the rumen to maintain sufficient levels of rumen ammonia for most efficient fermentation, so methods that increase levels of soluble protein in the diet (e.g., supplements, N fertilization, grazing management, legumes) may increase performance of cattle using these grasses. Avoiding frequent and severe grazing, maintaining adequate growth going into winter, and providing adequate rest between grazing periods will permit warm-season grass pastures to remain productive.

REFERENCES

Abrams, S.M., H. Hartadi, C.M. Chaves, J.E. Moore, and W.R. Ocumpaugh. 1981. Relationship of forage-evaluation techniques to the intake and digestibility of tropical grasses. p. 508–511. *In* Proc. XIV Int. Grassl. Congr., Lexington, KY. 15–24 June 1981. Westview Press, Boulder, CO.

Akin, D.E., J.E. Amos, F.E. Barton II, and D. Burdick. 1973. Rumen microbial degradation of grass tissue revealed by scanning electron microscopy. Agron. J. 65:825–828.

Akin, D.E., and D. Burdick. 1975. Percentage of tissue types in tropical and temperate grass leaf blades and degradation of tissues by organisms. Crop Sci. 15:661–668.

Anderson, B. 1986. Warm-season grasses: Facts and fantasy. p. 11–20. *In* Proc. Four State Grassl. Manage. Workshop. Warm Season Grasses: Facts and Fantasy, Maryville, MO. 7–9 July. Maryville, MO. Kansas State Univ., Manhattan.

Anderson, B. 1996a. Continuous vs rotational stocking of mixed warm-season grasses at three stocking rates. p. 110. *In* Proc. Am. Forage and Grassl. Coun., Vancouver, BC. 13–15 June 1996. Am. Forage and Grassland Council, Georgetown, TX.

Anderson, B. 1996b. Effectively grazing warm-season grasses. p. 22–31. *In* Native warm-season grass conference and expo. Iowa Forage and Grassl. Coun., Des Moines.

Anderson, B. 1999. Study duration can affect results and conclusions of grazing studies. p. 2. *In* Proc. of the 52nd Annual Meeting of the Soc. for Range Management, Omaha, NE. Soc. for Range Management, Denver, CO.

Anderson, S.J., T.J. Klopfenstein, and V.A. Wilkerson. 1988. Escape protein supplementation of yearling steers grazing smooth brome pastures. J. Anim. Sci. 66:237–242.

Blasi, D.A., J.K. Ward, T.J. Klopfenstein, and R.A. Britton. 1991. Escape protein for beef cows: III. Performance of lactating beef cows grazing smooth brome or big bluestem. J. Anim. Sci. 69:2294–2302.

Branson, F.A. 1953. Two new factors affecting resistance of grasses to grazing. J. Range Manage. 6:165–171.

Brown, R.H. 1978. A difference in N use efficiency in C_3 and C_4 plants and its implications in adaptation and evolution. Crop Sci. 18:93–98.

Burns, J.C., R.D. Mochrie, and D.H. Timothy. 1984. Steer performance from two perennial *Pennisetum* species, switchgrass, and a fescue: 'Coastal' bermudagrass system. Agron. J. 76:795–800.

Chapman, L.L., and A.E. Kretschmer. 1964. Effect of nitrogen fertilization on digestibility and feeding value of pangolagrass hay. Proc. Soil Crop Sci. Soc. Fla. 24:176–183.

Conard E.C., and D.C. Clanton. 1963. Cool-season, warm-season pastures needed. p. 11–13. Beef Cattle Progress Rep., Univ. of Nebraska, Lincoln.

Cuomo, G.J., and B.E. Anderson. 1996. Nitrogen fertilization and burning effects on rumen protein degradation and nutritive value of native grasses. Agron. J. 88:439–442.

Downes, R.W. 1969. Differences in transpiration rates between tropical and temperate grasses. Planta 88:261–273.

Duke, J.A. 1978. The quest for tolerant germplasm. p. 1–61. *In* G.A. Jung (ed.). Crop tolerance to suboptimal land conditions. ASA Spec. Publ. 32. ASA, Madison, WI.

Dwyer, D.D., and W.C. Elder. 1964. Grazing comparison of woodward sand bluestem and caddo switchgrass in Oklahoma. Oklahoma Agric. Exp. Stn. Bull. B-628. Oklahoma Agric. Exp. Stn., Stillwater.

Fairbairn, C.A., G.A. Jung, and L.D. Hoffman. 1985. Use of switchgrass to extend summer grazing in S.E. Pennsylvania. p. 9. *In* Agronomy abstracts. ASA, Madison, WI.

George, J.R., R.L Hintz, K.J. Moore, S.K. Barnhart, and D.R. Buxton. 1996. Steer response to rotational or continuous grazing on switchgrass and big bluestem pastures. p. 150–154. *In* Proc. Am. Forage and Grassl. Coun., Vancouver, British Columbia, Canada. 13–15 June 1996. Am. Forage and Grassland Council, Georgetown, TX.

Gerrish, J.R., P.R. Peterson, F.A. Martz, and R.E. Morrow. 1994. Impact of grazing management on the production and persistence of big bluestem pastures. p. 299–303. *In* Proc. Am. Forage and Grassland Council, Lancaster, PA. 6–10 Mar. 1994. Am. Forage and Grassland Council, Georgetown, TX.

Griffin, J.L., P.J. Wangsness, and G.A. Jung. 1980. Forage quality evaluation of two warm-season range grasses using laboratory and animal measurements. Agron. J. 72:951–956.

Grigsby, K.M., F.M. Rouquette, Jr., M.J. Florence, R.P. Gillespie, W.C. Ellis, and D.P. Hutcheson. 1988. Self-limiting supplemental protein for calves grazing bermudagrass pastures. Forage Research in Texas CPR-4593:16. Texas A&M Univ., College Station.

Hafley, J.L., B.E. Anderson, and T.J. Klopfenstein. 1993. Supplementation of growing cattle grazing warm-season grass with proteins of various ruminal degradabilities. J. Anim. Sci. 71:522–529.

Hunter, R.A., and B.D. Siebert. 1985. Utilization of low-quality roughage by *Bos indicus* cattle: 1. Rumen digestion. Br. J. Nutr. 53:637–648.

Kaiser, C.J., G.F. Cmarik, and D.B. Faulkner. 1986. How warm-season grasses perform in the southern Corn Belt. p. 23–26. *In* W.C. Sharp and T.N. Shiflet (ed.). Warm-season grasses: Balancing forage programs in the Northeast and southern Corn Belt. Soil Conserv. Soc. of Am. Ankeny, IA.

Kirch, B.H. 1995. Rumen escape protein and diet quality of cattle grazing smooth bromegrass, switchgrass, and big bluestem. Ph.D. diss. Univ. of Nebraska, Lincoln.

Klopfenstein, T., and M. Lewis. 1988. Optimizing beef production grazing cool-season and warm-season pastures. p. 6–10. *In* B. Anderson (ed.). Proc. Nebraska For. and Grassl. Council. Univ. of Nebraska, Lincoln. 7 Jan. 1988. Nebraska Forage and Grassland Council, Lincoln.

Krueger, C.R., and D.C. Curtis. 1979. Evaluation of big bluestem, indiangrass, sideoats grama, and switchgrass pastures with yearling steers. Agron. J. 71:480–482.

Ku, M.S.B., M.R. Schmitt, and G.E. Edwards. 1979. Quantitative determination of RuBP carboxylase-oxygenase in leaves of several C3 and C4 plants. J. Exp. Bot. 30:89–98.

Laetsch, W.M. 1974. The C4 syndrome: A structural analysis. Ann. Rev. Plant Physiol. 25:27–51.

Maposse, I.C., D.I. Bransby, and B.E. Gamble. 1996. Rotational and continuous stocking of steers grazing Alamo switchgrass in south Alabama. p. 24–28. *In* Proc. Am. Forage and Grassl. Coun. Vancouver, British Columbia, Canada. 13–15 June 1996. Am. Forage and Grassland Council, Georgetown, TX.

Maposse, I.C., D.I. Bransby, D.D. Kee, B.E. Gamble, and W. Gregory. 1995. Effect of stocking rate on stocker weight gains from Alamo switchgrass in Alabama. p. 24–27. *In* Proc. Am. Forage and Grassl. Coun., Lexington, KY. 12–14 Mar. 1995. Am. Forage and Grassland Council. Georgetown, TX.

Matches, A.G., G.B. Thompson, and F.A. Martz. 1975. Post-establishment harvesting and management systems of forages. p.111–135. *In* Proc. 1975 No Tillage For. Symp., Ohio State Univ., Columbus, OH.

Matches, A.G., F.A. Martz, D.P. Mowry, and S. Bell. 1982. Integration of subtropical perennial grasses into forage-animal systems. p. 150–151. *In* Agronomy abstracts. ASA, Madison, WI.

Miller, M.S., L.E. Moser, S.S. Waller, T.J. Klopfenstein, and B.H. Kirch. 1996. Immunofluorescence localization of RuBPCase in degraded C4 grass tissue. Crop Sci. 36:169–175.

Minson, D.J. 1981. Nutritional differences between tropical and temperate pastures. p. 143–157. *In* F.H.W. Morley (ed.). Grazing animals. Elsevier Sci. Publ. Co., Amsterdam.

Mitchell, R., L. Moser, B. Anderson, and S. Waller. 1994. Switchgrass and big bluestem for grazing and hay. NebGuide G94-1198. Coop. Ext. Univ. of Nebraska, Lincoln.

Moore, K.J., and D.R. Buxton. 1999. Fiber composition and digestion of warm-season grasses. p. 23–33. *In* K.J. Moore and B.E. Anderson (ed.) Native warm-season grasses: research trends and issues. CSSA Spec. Publ. 30. CSSA and ASA, Madison, WI.

Morgan, J.A., and R.H. Brown. 1979. Photosynthesis in grass species differing in carbon dioxide fixation pathways: II. A search for species with intermediate gas exchange and anatomical characteristics. Plant Physiol. 64:257–262.

Morris, R.J., R.H. Fox, and G.A. Jung. 1982. Growth, P uptake, and quality of warm- and cool-season grasses on a low available P soil. Agron. J. 74:125–129.

Morrow, R.E., D.J. Quinlan, M.S. Kerley, J.R. Gerrish, and F.A. Martz. 1994. Influence of grazing system on intake when cows graze big bluestem pastures. p.229–232. *In* Proc. Am. Forage and Grassl. Coun., Lancaster, PA. 6–10 Mar. 1994. Am. Forage and Grassland Council, Georgetown TX.

Mullahey, J.J., S.S. Waller, K.J. Moore, L.E. Moser, and T.J. Klopfenstein. 1992. In situ ruminal protein degradation of switchgrass and smooth bromegrass. Agron. J. 84:183–188.

Redfearn, D.D., and K.J. Hollingsworth-Jenkins. 1999. Escape and rumen degradable protein fractions in warm-season grasses. p. 3–21. *In* K.J. Moore and B.E. Anderson (ed.). Native warm-season grasses: research trends and issues. CSSA Spec. Publ. 30. CSSA and ASA, Madison, WI.

Redfearn, D.D., L.E. Moser, S.S. Waller, and T.J. Klopfenstein. 1995. Ruminal degradation of switchgrass, big bluestem, and smooth bromegrass leaf proteins. J. Anim. Sci. 73:598–605.

Reid, R.L., and G.A. Jung. 1985. Sward utilization in warm season-cool season grass grazing systems in the Northeast United States. p. 1111–1113. *In* Proc. XV Int. Grassl. Congr., Kyoto, Japan. 24–31 Aug. 1985. Science Council of Japan and the Japanese Soc. of Grassland Science, Nishi-nasuno, Tochigi-ken, Japan.

Reid, R.L., G.A. Jung, and W.V. Thayne. 1988. Relationships between nutritive quality and fiber components of cool season and warm season forages: a retrospective study. J. Anim. Sci. 66:1275–1291.

Rountree, B.H., A.G. Matches, and F.A. Martz. 1974. Season too long for you grass pasture? Crops Soils 26(7):7–10.

Smart, A.J., D.J. Undersander, and J.R. Keir. 1995. Forage growth and steer performance on Kentucky bluegrass vs. sequentially grazed Kentucky bluegrass-switchgrass. J. Prod. Agric. 8:97–101.

Smith, E.F., and C.E. Owensby. 1978. Intensive-early stocking and season-long stocking of Kansas flint hills range. J. Range Manage. 31:14–17.

Vaughn, D.R. 1997. Ruminal degradation of RuBPCase by beef cattle grazing switchgrass and big bluestem. M.S. thesis. Univ. of Nebraska, Lincoln.

Vona, L.C., G.A. Jung, R.L. Reid, and W.C. Sharp. 1984. Nutritive value of warm-season grass hays for beef cattle and sheep: digestibility, intake and mineral utilization. J. Anim. Sci. 59:1582–1593.

Ward, J.K. 1988. Optimizing beef production with cool- and warm-season pastures. p. 11–13. *In* B. Anderson (ed.). Proc. Nebraska For. and Grassl. Council. Univ. of Nebraska, Lincoln. 7 Jan. 1988. Nebraska Forage and Grassland Council, Lincoln.

Wedin, W.R., S. Barnhart, J. Russell, and W. Fruehling. 1980. Bromegrass and switchgrass grazing trials. p. 16–17. *In* Ann. Prog. Rep., Western Res. Ctr., Castana, IA. ORC 80-10. Iowa State Univ, Ames.

Wedin, W.F., and W. Fruehling. 1977. Smooth bromegrass vs. switchgrass for midsummer pasture. p. 9–11. *In* Ann. Prog. Rep., Western Res. Ctr., Castana, IA. ORC 77-10. Iowa State Univ, Ames.

Wilson, J.R., R.H. Brown, and W.R. Windham. 1983. Influence of leaf anatomy on the dry matter digestibility of C_3, C_4, and C_3/C_4 intermediate types of *Panicum* species. Crop Sci. 23:141–146.

11 Managing Weeds to Establish and Maintain Warm-Season Grasses[1]

Rob B. Mitchell and Carlton M. Britton
Texas Tech University
Lubbock, Texas

This chapter addresses some of our understanding of warm-season grass establishment and management. No attempt will be made to address all potential problems associated with the establishment and management of all warm-season grass species. We will present, however, information on native and introduced perennial species important to the Great Plains and Southern High Plains. During establishment, the management focus will be the control of grassy and broad-leaved herbaceous weeds. On established stands, we will concentrate on the management and control of cool-season grasses, and briefly address invasive woody plants. The general principles presented will probably have application to management aspects for all warm-season grass species and their associated problems.

The grasslands of the central USA were once among the most extensive and floristically diverse communities in North America (Risser et al., 1981). These grasslands were dominated by warm-season grasses such as big bluestem (*Andropogon gerardii* Vitman var. *gerardii* Vitman), indiangrass [*Sorghastrum nutans* (L.) Nash], little bluestem [*Schizachyrium scoparium* (Michx.) Nash], switchgrass (*Panicum virgatum* L.), and sideoats grama [*Bouteloua curtipendula* (Michx.) Torr.]. Prior to permanent settlement, 283 million ha of native prairie existed west of Ohio, with about 85% of Iowa covered by tallgrass prairie (Heath & Kaiser, 1985). Today, <3% of the original 100 million ha of Great Plains tallgrass prairie remains (Smith, 1992). Remaining grasslands are often in degraded condition because of the encroachment of introduced cool-season grasses such as Kentucky bluegrass (*Poa pratensis* L.) and smooth bromegrass (*Bromus inermis* Leyss.). These cool-season invaders reduce summer forage quantity and quality, and reduce native species diversity. Grasslands with a history of poor grazing management or where fire has been excluded are particularly prone to cool-season grass invasion (Masters & Vogel, 1989).

Large tracts of grassland dominated by native warm-season grasses remain intact in Nebraska, Kansas, Oklahoma, and Texas; however, a majority of the grasslands in the central USA have been converted to cropland because climate,

---

[1] Contribution number T-9-838 of the College of Agricultural Sciences and Natural Resources, and Fire Ecology Center Technical Paper 6, Texas Tech University.

Copyright © 2000. Crop Science Society of America and American Society of Agronomy, 677 S. Segoe Rd., Madison, WI 53711, USA. *Native Warm-Season Grasses: Research Trends and Issues.* CSSA Special Publication no. 30.

soil, and topography are ideally suited to annual crop production. Cultivated monocultures of corn (*Zea mays* L.), sorghum [*Sorghum bicolor* (L.) Moench], soybean [*Glycine max* (L.) Merr.], wheat (*Triticum aestivum* L.), and cotton (*Gossypium hirsutum* L.) now dominate the landscape. In Iowa, Minnesota, Missouri, Nebraska, North Dakota, and South Dakota, nearly 8 million ha of cropland are in land capability classes IV through VIII (USDA-NRCS, 1994). Land in these capability classes is usually best suited to perennial forage production rather than annual crops. Consequently, millions of hectares of cultivated land in the Great Plains and the Southern High Plains are candidates for conversion from cropland back to perennial grasses.

Monocultures or mixtures of warm-season grasses such as switchgrass, big bluestem, indiangrass, little bluestem, and weeping lovegrass [*Eragrostis curvula* (Schrad.) Nees] have been seeded throughout the Great Plains and Southern High Plains as part of the Conservation Reserve Program (CRP) contained in the 1985 Farm Bill. These species are important forage sources during the summer months when cool-season grasses become less productive. With the expiration of contracts from the original CRP, millions of hectars of perennial grass will require more intensive management.

WARM-SEASON GRASS ESTABLISHMENT

Successful establishment of warm-season grasses requires selection of an appropriate species and cultivar for the site capability and region, proper seedbed preparation, proper planting depth, and weed control following seeding. Seedling establishment occurs when a plant reaches a stage of development where leaf area and root development can maintain continued plant growth (Wolf et al., 1996). Stand establishment occurs when the period of high seedling mortality is over and an acceptable number of plants have developed adequate canopies and root systems to maintain their position in the community (Booth & Haferkamp, 1995). Seedling morphology is an important consideration covered in a previous chapter of this publication. Due to variability in sites, conditions, and budgets, the following section discusses important general considerations in seedbed preparation and seeding. Consult the references and contact local experts for more site-specific information.

Seeding Considerations

In native rangelands, the decision to artificially revegetate an area can be difficult. In many cases, the natural revegetation process, allowing secondary succession to progress, may be an appropriate solution that minimizes capital expenditures. For improved pastures, the most important decision in the seeding process is selection of an appropriate species and cultivar, or mixture of species and cultivars, to meet management objectives. After regionally adapted species and cultivars have been identified, it is necessary to determine if the selection is well adapted to the site, specifically soil texture, water-holding capacity, pH, and aspect. Contact the USDA-Natural Resources Conservation Service, USDA-Cooperative

Extension Service, or university personnel for species and cultivars best adapted to your region and soil conditions.

Seeding method and soil conditions will determine the required seedbed preparation. Vallentine (1989) identified six important characteristics of an ideal seedbed: (i) very firm below the seeding depth, (ii) well pulverized and mellow on top, (iii) not cloddy nor puddled, (iv) free from live, resident plant competition, (v) free from seed of competitive species, and (vi) has moderate amounts of mulch or plant residue on the soil surface. The method used for seedbed preparation depends on the kind and amount of vegetation remaining on the site, the susceptibility of the site to wind and water erosion, salinity, stoniness, texture, depth, accessibility, obstructions, and cost limitations (Vallentine, 1989). Choices for seedbed preparation generally include mechanical tillage, which may be combined with preparatory crop planting and herbicide application to control weeds. Numerous implements are available to provide proper seedbed preparation in variable soil conditions, and numerous seeding implements are available depending on conditions. Regardless of which seeding implement is selected, the proper planting depth for most warm-season grasses ranges from 0.5 cm for small seeds such as weeping lovegrass to 1.5 cm for relatively larger seeds such as switchgrass.

Adequate plant density is critical for successful stand establishment, so proper seeding rate is important. Launchbaugh (1966) surveyed >3000 grass plantings of numerous species in the Great Plains and classified these plantings as good stands if 10 or more seedlings m^{-2} were established. Launchbaugh and Owensby (1970) evaluated five native warm-season grasses and determined 10 to 20 plants m^{-2} during the establishment year provided good stands. Vogel (1987) determined seeding rates in eastern Nebraska of >200 pure live seed (PLS) m^{-2} were not necessary to obtain adequate stands of big bluestem and switchgrass. He suggested seeding rates of 100 PLS m^{-2} may be adequate for conservation plantings that do not necessitate grazing the year after planting.

Weeping lovegrass has been used extensively on sandy sites in Texas and Oklahoma where annual precipitation exceeds 38 cm (Dahl & Cotter, 1984; Marietta & Britton, 1989). Weeping lovegrass generally has good seedling vigor, but establishment can be difficult in unprepared seedbeds with competition from established vegetation (Dahl & Cotter, 1984). The best seedbed preparation for weeping lovegrass is a clean, firm seedbed, much like for small grains. Sandbur (*Cenchrus incertus* M.A. Curtis) often provides the greatest competition for weeping lovegrass establishment, and has caused stand failure (Marietta & Britton, 1989). The best way to alleviate sandbur competition is moldboard plowing rather than discing during site preparation to bury the sandbur seed (Dahl & Cotter, 1984; Marietta & Britton, 1989).

Weed Control

Competition or interference from weeds often limits stand establishment of perennial warm-season grasses or may cause complete stand failure (Martin et al., 1982; Masters, 1995). Alleviation of weed interference by using appropriate cultural practices or applying pre- and post-emergence herbicides enhances grass establishment.

Atrazine [6-chloro-N-ethyl-N'-(1-methylethyl)-1,3,5-triazine-2,4-diamine] applied at planting has dramatically improved establishment of selected atrazine-tolerant cultivars of big bluestem and switchgrass in Nebraska (Martin et al., 1982). Sand bluestem [*Andropogon gerardii* var. *paucipilus* (Nash) Fern.] establishment on sandy soils in Nebraska was improved by atrazine application at planting, but provided no benefit to prairie sandreed (*Calamovilfa longifolia* Hook.) establishment on these sandy soils (Masters et al., 1990). They recommended not applying atrazine at planting when prairie sandreed is a component of the grass seed mixture seeded on sandy soils. Apparently, prairie sandreed seedlings are intermediate in tolerance to atrazine (Bahler et al., 1984). Significant weed interference, however, should be expected on these sandy sites because most areas requiring revegetation will be abandoned or marginal cropland in an early seral stage. Current label restrictions for atrazine limit use to land enrolled in the CRP.

Annual grasses such as crabgrass [*Digitaria sanguinalis* (L.) Scop.], fall panicum (*Panicum dichotomiflorum* Michx.), green foxtail [*Setaria viridis* (L.) Beauv.], yellow foxtail [*Setaria glauca* (L.) Beauv.], and barnyardgrass [*Echinochloa crusgalli* (L.) Beauv.] often pose the most significant threat to warm-season grass establishment. Metolachlor [2-chloro-N-(2-ethyl-6-methylphenyl)-N-(2-methoxy-1-methylethyl)acetamide] has provided better control of these warm-season annual grasses than atrazine. Metolachlor applied at planting improved establishment of big bluestem and sand bluestem cultivars when applied alone or in combination with atrazine where fall panicum, green foxtail, yellow foxtail, velvetleaf (*Abutilon theophrasti* Medic.), and redroot pigweed (*Amaranthus retroflexus* L.) provided most of the weed pressure. Big bluestem and sand bluestem yield increased by at least 30% when metolachlor and atrazine were applied preemergence the year of planting compared to no herbicide application. Yield and stand frequency were similar when either herbicide was applied, indicating metolachlor is a suitable replacement for atrazine to improve big bluestem and sand bluestem establishment.

MANAGING NATIVE AND ESTABLISHED STANDS

Methods of manipulating grassland vegetation to improve forage production have included prescribed burning, fertilization, and herbicide application. Factors influencing the response of grassland vegetation to these treatments include plant community successional status (Gillen et al., 1987), plant water status (Hake et al., 1984; Shiflet & Dietz, 1974), soil water content (Anderson, 1965), time of spring burning (Anderson, 1965; Mitchell et al., 1996; Towne & Owensby, 1984), and time of fertilizer application (Rehm et al., 1976).

Masters et al. (1996) proposed a successional model for Great Plains grasslands (Fig. 11–1). Retrogression is initiated by the combined effects of factors such as overgrazing, fire exclusion, conversion to cropland, and the introduction of invasive exotic species. Retrogression leads to a steady state of increased encroachment of invasive exotic species and reduced native species diversity. Managing succession to increase native species diversity and productivity can be accomplished by burning, fertilizing, applying herbicides, planting native species, or using proper

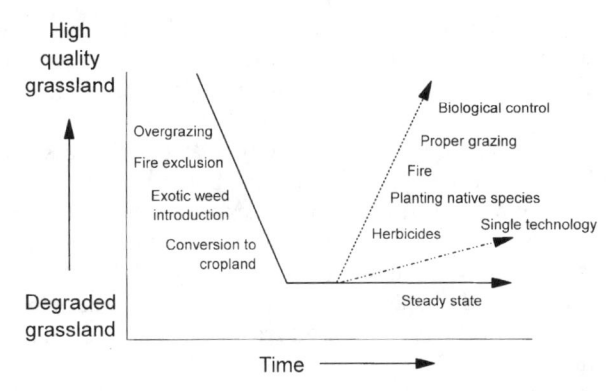

Fig. 11–1. Successional process model for Great Plains grasslands (from Masters et al., 1996).

grazing management in complimentary sequences or combinations. Dependence on a single practice, however, may result in a slow rate of stand recovery.

Fire

Fire was an important factor in the formation and maintenance of grassland ecosystems. Grasslands developed under extreme fluctuations in temperature and precipitation, with extended periods of dry weather followed by erratic storms that promoted the fire regime responsible for the central USA grasslands. Primary fire ignition sources were lightning and indigenous people (Humphrey, 1958).

Fire restricted woody vegetation encroachment into grassland ecosystems (Bragg & Hulbert, 1976), prevented excessive mulch accumulation on the soil surface (Vogl, 1974), and accelerated decomposition and nutrient cycling (Sauer, 1944). Wildfires in grasslands occurred during all four seasons (Wright, 1974), but fire probability was highest in summer when increasing vegetation maturity and lightning occurrences provided the dry fuel and ignition source (Bragg, 1982). Wildfires often occurred in late summer or fall when winds and air temperatures were high and relative humidity was low (Wright & Bailey, 1982). Late summer or fall fires left the soil surface exposed and unprotected from wind, rapid freezing and thawing, and drying (Weaver, 1954) and could shift species composition away from warm-season dominants to undesirable cool-season grasses (Ewing & Engle, 1988). Late fall and winter wildfires also occurred, but the probability of fires being naturally ignited was low because of the low probability of lightning storms (Bragg, 1982). Probability of fire ignition during spring and early summer was great because of the increased incidence of lightning storms, presence of dead fine fuel, and dormancy of herbaceous vegetation.

Prescribed Fire

Prescribed fires are systematically planned and carried out to meet specific management objectives (Scifres, 1980; Wright & Bailey, 1982). Special considerations are given to wind speed and direction, soil moisture, fuel load, ambient air temperature, relative humidity, fuel moisture and volatility, and growth stage of plants (Wright, 1974). In contrast, wildfires are unplanned and occur as a result of lightning or human negligence and can have an undesirable effect on plant communities.

Grassland response to burning is directly related to the season or timing of the burn. Prescribed fires in the Great Plains and Southern High Plains are usually applied in the spring when warm-season grasses are dormant, soil moisture is adequate to support plant growth, weather patterns are predictable, and removal of high quality forage is minimized (Wright, 1974; Bragg, 1982). Late fall and winter fires generally are not desirable because soil exposure to desiccation is increased, which inhibits plant growth, promotes soil erosion, and reduces water infiltration. Although summer fires likely represent the natural fire regime, our understanding of prescriptions for safely conducting summer burns and predicting fire behavior under these extremes in air temperature and relative humidity is poor. Additionally, the loss of available forage and potential invasion by nondesirable species, primarily exotics, is not understood. Consequently, we recommend summer burning be restricted to experimental burning for research purposes only until more information is available.

Prescribed fire can meet many management objectives simultaneously in warm-season grasslands (Wright & Bailey, 1982). A single prescribed fire can enhance stand vigor (Hadley & Kieckhefer, 1963; Knapp & Seastedt, 1986; Weaver & Tomanek, 1951), improve animal performance (Allen et al., 1976; Woolfolk et al., 1975), reduce cool-season invaders (Anderson et al., 1970; Ehrenreich, 1959; McMurphy & Anderson, 1965; Old, 1969), retain species diversity (Kucera, 1970), increase dry matter production (Curtis & Partch, 1948; Ehrenreich & Aikman, 1957; Mitchell et al., 1994; Rice & Parenti, 1978; Robocker & Miller, 1955), increase seed production (Cornelius, 1950; Curtis & Partch, 1950; Masters et al., 1993), increase forage quality (Mitchell et al., 1994), and suppress woody vegetation (Bragg & Hulbert, 1976; Wright, 1974). For more information about using prescribed fire in many grassland systems, see Wright and Bailey (1982).

Dry Matter Production

Burning rarely results in the death of warm-season perennial grass plants; however, no consistent responses in dry matter production have been observed for any warm-season grasses in all environments. Plant response to fire depends on plant morphology, stage of plant development at burning (Scifres, 1980), and plant water status (Hake et al., 1984; Shiflet & Dietz, 1974). The rhizomatous habit of big bluestem, switchgrass, and indiangrass may be an important morphologic characteristic determining their response to fire (Ewing & Engle, 1988). Period of crown burning is reduced in rhizomatous species compared with cespitose (bunchgrass) species, minimizing injury to meristematic tissue in rhizomatous species. In the cespitose species weeping lovegrass, we have observed smoldering in the center of the plant crown for 70 min after passage of the fire front. Rhizomatous species have the potential for increased tiller development following removal of the dormant canopy. Studies in the tallgrass prairie have focused on burning during warm-season grass dormancy. Towne and Owensby (1984) reported burning tallgrass prairie late in the spring increased rhizomatous warm-season grasses, but decreased bunchgrasses, cool-season grasses, sedges, and forbs. The cool-season grasses, sedges, and forbs had elevated growing points at burning, which made them susceptible to fire injury.

Availability of soil water may limit dry matter production following spring burning. Studies in tallgrass prairies have reported reductions in dry matter yield following spring burning (Anderson, 1965; Ehrenreich & Aikman, 1963; McMurphy & Anderson, 1965). These decreases in dry matter production were attributed primarily to lack of water infiltration and loss of soil moisture. Hanks and Anderson (1957) reported soil moisture reductions were related to decreased water infiltration following litter removal. Removing litter exposes the soil surface to wind and direct sunlight, which increases evaporative water loss.

Spring burning native warm-season grass swards when soil water was adequate resulted in increased dry matter production (Abrams et al., 1986; Curtis & Partch, 1950; Ehrenreich & Aikman, 1957; Mitchell et al., 1994). Positive grassland responses to fire when soil water is adequate result from removal of litter, which increases insolation and reduces shading. These factors cause the soil to warm earlier and increase microbial activity, which increases nutrient availability to the plants. These factors may cause warm-season grasses to initiate growth 7 to 10 d earlier compared with nonburned sites (Kucera & Ehrenreich, 1962).

Soil water content of native grasslands is affected by spring burning date (Anderson, 1965). Aldous (1934) determined that soil water content of the top 1 m of the soil profile in bluestem range was reduced by spring burning. Early spring burning (March 20) caused the greatest reduction in soil moisture, followed by midspring (April 10), and late spring burning (May 1; Anderson, 1965). Early spring burning left the soil surface exposed to water loss by evaporation, runoff, and surface erosion before regrowth began (Anderson, 1965; Anderson et al., 1970). Late spring burning protected the soil surface from these phenomena until later in the growing season when plant growth was more rapid. Date of burning in relation to growth cycles of species determined species response (Anderson et al., 1970). Anderson (1965) stated, "The earlier one burns, the less herbage he will have for harvest by livestock."

Fire mineralizes organic N into inorganic forms and reduces the total N in the system, but increases amount of N present in forms that are available for plant uptake (Christensen, 1973; St. John & Rundel, 1976; Hobbs et al., 1991). Loss of total N results from the pyrolysis of the plant amino acid N that is converted to NH_3, N_2, or various oxides (NO) (St. John & Rundel, 1976). These forms of N are converted by soil bacteria to plant available NO_3^- or NH_4^+. The amount of total N lost during burning increases as fire temperature increases (St. John & Rundel, 1976; Hobbs et al., 1991). Maximum fire temperatures at the mineral soil surface varied from 83 to 680°C for fine fuels ranging from 1690 to 7870 kg ha^{-1} (Stinson & Wright, 1969).

Fire increases soil temperature by direct heating from the fire front and by increasing solar insolation. Soil surface temperatures as high as 260°C were recorded by Gay and Dwyer (1965), but temperatures at the rhizome level (1.27 cm below the soil surface) were not sufficient to cause plant injury. Soil temperature increases below the surface by direct heating from the fire front are minimal and did not generally affect aboveground production (Hulbert, 1988). Soil temperature increases from increased solar insolation are more important and long-term (Kucera & Ehrenreich, 1962; Old, 1969; Weaver & Rowland, 1952). Soil temperatures were 1.0 to 2.2°C higher throughout the growing season for burned areas than for nonburned areas (Hensel, 1923). Soil surface temperatures about 1 mo after burning were 3.9°C higher on burned plots than nonburned plots, and subsoil temperatures (10 cm) were 2.8°C higher on burned plots than nonburned plots (Hulbert, 1988). Increased solar radiation absorption on blackened soils and decreased shading from mulch removal caused the higher temperatures. Increasing soil temperature increased N mineralization (Hobbs & Schimel, 1984; St. John & Rundel, 1976; Stanford et al., 1973) and NO_3^- production by soil microbes (Sharrow & Wright, 1977).

Removal of accumulated plant biomass by fire alters the quantity and quality of light reaching the plant. Anderson et al. (1970) reported that prairies excluded from grazing and fire soon began to deteriorate due to the accumulation of litter that depressed herbage yields and species diversity. Grazing and/or fire was necessary to prevent accumulation of excessive litter. Big bluestem, a tallgrass prairie dominant, responds favorably to increased irradiance after spring burning by increasing photosynthetic rate, stomatal conductance, leaf thickness (Knapp, 1985), and tiller density (Ewing & Engle, 1988). Nonburned plants had decreased stomatal density (−7%) and pore length (−22%), totaling a 41% decrease in stomatal conductance (Knapp & Gilliam, 1985). Additionally, nonburned plants had decreased leaf thickness (−18%), specific leaf mass (−34%), and bundle sheath–vascular complex (−23%), resulting in decreased leaf biomass production.

Little information is available concerning the impacts of late growing-season fires on tallgrass prairie vegetation. Engle et al. (1998) evaluated mid-successional tallgrass prairie response to several fire frequencies during August, September, and October. They concluded late growing-season fires in mid-successional tallgrass prairies did not reduce total production, but reduced forage grass production and increased forb production, which may improve bobwhite quail (*Colinus virginianus* L.) habitat. Although they did not recommend late growing-season fires as a management practice to induce succession, it does appear likely late growing-season fire can induce succession of disturbed tallgrass prairies in a short time period. If maintenance of woody cover is a desire, however, avoid late growing-season fires.

Most burning research on native warm-season grasses has been conducted on mixed-species stands, so interactions among species may mask fire effects; however, Cuomo et al. (1996) evaluated the impact of spring burning date on monocultures of switchgrass, big bluestem, and indiangrass in a 3-yr study in eastern Nebraska. Annual burning on all dates consistently reduced dry matter production in all species, but reductions were greatest on areas burned in May. When averaged across all burn dates and years, burning reduced yield of switchgrass by 41%, indiangrass by 59%, and big bluestem by 30%, which is contrary to research on annual burning in tallgrass prairies (Towne & Owensby, 1984). Cuomo et al. (1996) concluded burning in consecutive years was detrimental to the productivity of these species grown in monocultures. Due to yield reductions associated with annual burning in monocultures of these species, they recommend burning to meet specific management objectives, such as removing litter or manipulating species composition.

Most burning research on introduced warm-season perennial grasses has been conducted on monocultures. Weeping lovegrass is an excellent example of an introduced warm-season perennial grass that is well suited to management as a monoculture. In the Southern High Plains of Texas, >400 000 ha of weeping lovegrass have been established on sandy sites since 1985. Dahl and Cotter (1984) indicated the primary objective for burning weeping lovegrass may be to remove decadent material and make forage more available, not necessarily increase yield. They recommended burning weeping lovegrass just prior to spring emergence to minimize yield reduction and limit the possibility of wind erosion on the sandy sites. Additionally, burning promotes new tiller development in the dead plant center and around the plant perimeter. McFarland and Mitchell (1997) reported spring burning increased weeping lovegrass tiller density. Average tiller density on non-burned and burned areas was 1070 and 1730 m^{-2}, respectively. Mullins et al. (1997) reported spring burning decreased weeping lovegrass standing green biomass during dry portions of the growing season, but increased production following periods of near-normal precipitation. Consequently, weeping lovegrass dry matter production response to burning is closely related to soil water content and growing conditions following burning.

Forage Quality

Prescribed burning can improve palatability and nutritional value of warm-season grasses. Removal of accumulated dead plant biomass improves availability of high quality forage by removing physical barriers to grazing. Steer average daily gains during the summer were higher on burned areas than on nonburned areas (Rao et al., 1973; Woolfolk et al., 1975). Greatest animal gains were observed on areas burned in late spring (Anderson et al., 1970). Smith et al. (1960) reported that burning increased apparent digestibility of forage from Kansas bluestem pastures. A trend of increased protein digestibility was apparent following burning, but increases were statistically significant in only one of four trials.

Vogl (1965) determined spring burning increased water content of forbs and grasses in regrowth following burning. This increased succulence of grasses and forbs makes them more desirable and palatable to herbivores, resulting in increased utilization. Allen et al. (1976) found late spring burning decreased dry matter (DM), crude fiber, cell wall constituents, cellulose, and lignin, and increased crude protein (CP), ether extract, N-free extract, and ash of big and little bluestem. The CP concentration ranged from 17.7% in May to 2.9% in November, but remained relatively constant during August, September, and October (4.2 to 4.5%).

Mitchell et al. (1994) reported spring burning date impacted big bluestem in vitro dry matter disappearance (IVDMD), CP, and neutral detergent fiber (NDF) concentration. Big bluestem IVDMD in June was greatest from areas burned in late spring, in July was greatest from areas burned in early spring, and in August was greatest from areas burned in mid spring. Big bluestem IVDMD generally was lowest on nonburned areas. Big bluestem CP in June and July was greatest for areas burned in late spring, but by August CP was greater in nonburned areas than in any burn date. Big bluestem NDF was lower in June following burning in early, mid, or late spring, but responses in July and August were variable. A management scenario in which big bluestem-dominated pastures could be managed in a burning and grazing rotation during the same grazing season to take advantage of the forage quality improvements provided by different burning dates was presented.

Seed Production

Favorable responses of big bluestem to spring burning have included increased flowering stalk density and seed production (Benning & Bragg, 1993; Cornelius, 1950; Curtis & Partch, 1948; Curtis & Partch, 1950; Old, 1969; Masters & Vogel, 1989; Masters et al., 1993; Rice & Parenti, 1978). Increases in flowering stalk density and seed production were greater with late-spring burning than with early-spring burning (Benning & Bragg, 1993; Henderson et al., 1983; Masters et al., 1993; Olson, 1986). Increased flowering stalk density resulted from litter removal (Curtis & Partch, 1950) that released labile nutrients and improved the quantity and/or quality of light or removed the apical dominance that stimulated tillering from the axillary buds of the rhizomatous species (Risser et al., 1981).

Burning increased big bluestem reproductive stem density >20-fold when compared with no burning (Curtis & Partch, 1948), and six-fold in another study (Curtis & Partch, 1950). They concluded that height and number of flowering stalks were increased on the burned plots by the removal of the insulative layer of plant debris that permitted the plants to initiate growth early. Ehrenreich (1959) and Ehrenreich and Aikman (1963) found that early spring burning in Iowa increased flowering in most native warm-season grasses. The warm-season grasses initiated growth 2 to 3 wk earlier on burned plots and produced more and taller flowering stalks than nonburned plots. They attributed the increases to higher soil temperatures resulting from the removal of accumulated litter by the fire (Ehrenreich, 1959). In an Illinois prairie, the number of flowering stalks of big bluestem and indiangrass were greatest on areas that had been burned annually in three previous years (Hadley & Kieckhefer, 1963). Cessation of burning for 1 yr resulted in a decrease in flowering stalk density. A 10-fold increase in flowering stalk density of burned plots compared to nonburned plots also was reported in Illinois (Old, 1969). Clipping and removing the vegetation also increased flowering stalk density, but only about one-half as much as burning. Increased flowering induced by burning was attributed to the removal of litter from around the growing points of the grasses and reduction in the competition from cool-season invaders.

Benning and Bragg (1993) evaluated the importance of the timing of spring burning on big bluestem in reestablished tallgrass prairies in eastern Nebraska. Flowering stem height and flowering stem number increased on areas burned at 4-d intervals after 12 May; however, plants on areas burned at 4-d intervals prior to 12 May did not exhibit increased flowering stem height or number. This study reiterated the importance of timing fire application to meet specific management objectives. Additionally, interpreting results from fire studies conducted during broadly defined treatments such as spring burning should be done with care, and reporting such results should include some information on plant phenology at burning.

Masters et al. (1993) found flowering stalk density of big bluestem and indiangrass was affected by spring burning date. Burning in late spring consistently increased big bluestem reproductive stem density. Indiangrass response was more variable, with reproductive stem density being similar for no burning and late-spring burning in one environment, and all burning dates being greater than no burning in another environment. The combination of spring burning and fertilization increased indiangrass reproductive stem density by as much as five-fold when compared with no burning or fertilization in two environments. Germinable seed of big bluestem increased by at least 140% compared with no burning or burning in early or mid spring; however, maximum germinable seed was only 125 seeds m^{-2}.

Prairies are not reliable sources of seed in most years because of plant competition and adverse weather conditions such as high temperature and unreliable precipitation during flower initiation. Cornelius (1950) conducted a comprehensive study comparing seed yields from commercial seed production fields and native prairies. The native prairie seed yields were much more variable than seed yields from commercial seed production fields. Big bluestem in native prairies produced quantities of seed considered harvestable in only 3 of 9 yr, whereas cultivated stands produced harvestable quantities each year during the same period. Warm-season grass seed yields increases following late-spring burning and fertilization because fire reduced weed competition and fertilizer stimulated the growth of the grass. Number of germinable seed increased >600% following a combination of spring burning, fertilization, and atrazine application (Masters & Vogel, 1989).

Herbicides

Herbicides have the potential to selectively control problem species in warm-season grasslands. Historically, producers in the Great Plains have applied 2,4-D [(2,4-dichlorophenoxy)acetic acid], dicamba (3,6-dichloro-2-methoxybenzoic acid), and picloram (4-amino3,5,6-trichloro-2-pyridinecarboxylic acid) to control musk thistle (*Carduus nutans* L.) and leafy spurge (*Euphorbia esula* L.). In Kansas, spring application of 2.24 kg active ingredient (a.i.) ha^{-1} of 2,4-D controlled 99% of musk thistle plants, whereas fall application controlled 89% of plants (Fick & Peterson, 1995). Application of picloram, or a combination of 2,4-D and dicamba provided similar results, but spraying in the rosette stage prior to bolting was critical for all herbicides. In Nebraska, application of 4.7 to 9.4 l ha^{-1} of picloram has been reported to give 70 to 90% leafy spurge control the first year after treatment, but declined significantly over time without retreatment (Moomaw et al., 1989). Most of the research in warm-season grasslands of the Great Plains has focused on atrazine and glyphosate [N-(phosphonomethyl)glycine] application to selectively control cool-season invaders and improve the productivity of warm-season grasses in degraded tallgrass prairies (Morrow et al., 1977; Samson & Moser, 1982; Waller & Schmidt, 1983). Although atrazine is no longer labeled for rangeland use, the results of previous research are important to evaluate potential benefits associated with selective control.

Atrazine has selectively controlled cool-season grasses and forbs in warm-season grasslands (Houston, 1977; Morrow et al., 1977; Peterson et al., 1983; Rehm, 1984; Samson & Moser, 1982; Waller & Schmidt, 1983). Species most susceptible to atrazine include Kentucky bluegrass, Canada bluegrass (*Poa compressa* L.), smooth bromegrass, downy bromegrass (*Bromus tectorum* L.), green foxtail, annual bromes (*Bromus* sp.), and Russian thistle (*Salsola kali* L. var. *tenuifolia*

Tausch). Although seedlings of some native warm-season grasses are not atrazine-tolerant, most established warm-season grass plants are not damaged by atrazine.

Several studies have reported increases in warm-season grass forage yield, quality, and seed stalk density resulting from atrazine application. Waller and Schmidt (1983) found that 2.24 kg ha^{-1} of atrazine applied in April increased big bluestem herbage yield. They attributed the increase in big bluestem herbage yield to herbicide-induced suppression of the cool-season grasses. Atrazine reduced Kentucky bluegrass by 96% and smooth bromegrass by 91% at the end of the first growing season after application. Atrazine increased total forage production and shifted the period of optimum forage quality from spring and fall to summer and may have enhanced the drought resistance of big bluestem. Rehm (1984) reported atrazine application increased warm-season grass yields in years of adequate precipitation, but had no effect on warm-season grass yields in dry years. He stated that atrazine applied in early April eliminated the growth of smooth bromegrass and weeds, making the warm-season grasses more competitive for nutrients and moisture. Rehm (1984) found atrazine application had no significant effect on CP concentration, which does not agree with Houston and Van Der Sluijs (1975), who found atrazine increased CP by 21% over untreated range in Colorado. Rehm (1984) attributed this lack of agreement to differences in species evaluated. In California, Kay (1971) reported sub-herbicidal levels of atrazine increased the quantity and quality of range forage. Application of atrazine in Oklahoma increased the quantity (Baker & Powell, 1978; Engle et al., 1990; Gillen et al., 1987; Masters et al., 1992) and quality of tallgrass prairie forage (Baker & Powell, 1978). Masters and Vogel (1989) reported atrazine increased flowering stalk density at a remnant prairie site in eastern Nebraska.

Glyphosate, a nonselective post-emergent herbicide, is an effective herbicide for controlling invasive cool-season grasses in warm-season grasslands (Waller & Schmidt, 1983). Glyphosate can be applied in spring or fall when warm-season grasses are dormant and cool-season grasses are actively growing and susceptible to injury. Prior to treatment of an overgrazed tallgrass prairie in eastern Nebraska, Kentucky bluegrass and smooth bromegrass comprised 39 and 46%, respectively, of total vegetation (Waller & Schmidt, 1983). A single application of glyphosate in late spring stimulated remnant warm-season grasses and reduced Kentucky bluegrass and smooth bromegrass by 96 and 98%, respectively, in August the year of application. Big bluestem herbage yield was increased by 81% on areas treated with glyphosate when compared with nontreated areas. Additionally, big bluestem yield was not different on areas treated with glyphosate or atrazine. Glyphosate application increased total forage production and shifted the period of optimum forage quality and quantity from spring and fall to summer.

Treatment Combinations

Few studies have evaluated the combined effect of spring burning, fertilization, and/or herbicide application on grasslands (Gillen et al., 1987; Masters et al., 1992; Mitchell et al., 1996). Woolfolk et al. (1975) in the Kansas Flint Hills reported that the mean CP concentration of steer (*Bos taurus*) diets was higher on burned and fertilized pastures than on any other treatment. A combination of burning and N fertilization produced the highest average daily gains.

Rehm (1984) reported increased DM production of warm-season tallgrasses with the application of N and P fertilizer and atrazine. High variability of forage yields occurred due to fluctuations in rainfall from year to year. He showed in this 6-yr study the positive response of warm-season grasses to repeated applications of fertilizer in combination with atrazine. Gay and Dwyer (1965) reported herbage yields were not affected by burning alone, but with addition of N fertilizer, production increased from 3460 kg ha^{-1} on nontreated areas to 5030 kg ha^{-1}. The combined treatment of burning and fertilization also increased herbage yields by 54% over the nonburned areas treated with 112 kg ha^{-1} of N. Graves and McMurphy (1969) in central Oklahoma reported N fertilizer had the greatest effect on forage production. Spring burning decreased prairie threeawn (*Aristida oligantha* Michx.), and improved production of high quality warm-season grasses.

Owensby and Smith (1979) in the Kansas Flint Hills also combined spring burning and N fertilization. They noted big bluestem and indiangrass basal cover increased following burning and fertilization, but decreased on the nonburned pastures with the same fertilization rates. They also reported basal cover of Kentucky bluegrass increased as fertilizer rate increased on nonburned pastures. They concluded that burning and fertilization was more effective at maintaining good quality range than fertilizing alone.

Gillen et al. (1987) working in Oklahoma tallgrass prairie took range improvement one step further by including atrazine application with spring burning and fertilization. Atrazine and burning interacted on most dates to control annual grasses. But since both controlled annual grasses so effectively, little potential remained for an additive effect of treatment combinations. Atrazine generally reduced forb production and increased perennial grass production. Nitrogen interacted with atrazine and burning to affect forb production in August of both years. Nitrogen alone increased forb production, but burning and atrazine negated the effect. Nitrogen and burning did not interact positively to increase perennial grass production. No positive interaction existed between N and atrazine, which agreed with Baker and Powell (1978), but not with Rehm (1984). Gillen et al. (1987) concluded atrazine and spring burning were similar as weed control treatments, but atrazine tended to have a more positive effect on perennial grasses than spring burning.

Masters et al. (1992) reported standing crop increases of dominant warm-season grasses by 50 to 127% following spring burning combined with fertilization. Forb standing crop was reduced by atrazine or burning followed by atrazine application in four of seven grassland environments studied. This is consistent with findings from Oklahoma (Gillen et al., 1987). Masters et al. (1992) determined the primary utility of atrazine application is to renovate degraded prairie communities dominated by cool-season grasses, not to improve the productivity of tallgrass prairies in good to excellent condition.

Mitchell et al. (1996) evaluated the impacts of prescribed burning date, fertilization, and atrazine application on tallgrass prairie vegetation in eastern Nebraska. Burning in late spring maintained big bluestem accumulated standing crop through July, whereas no burning, and burning in early and mid spring resulted in declines during July. Burning in late spring reduced prairie dropseed [*Sporobolus het-*

erolepis (A. Gray) A. Gray] and tall dropseed [*Sporobolus asper* (Michx.) Kunth.], species that can dominate degraded sites. Late-spring burning in one environment reduced cool-season grasses by 86% when compared with nonburned areas. Fertilization increased big bluestem in June and July, but had little effect in August. Burning date combined with atrazine impacted indiangrass production, but specific treatment responses were variable. For example, mid-spring burning combined with atrazine reduced indiangrass standing crop by 560 kg ha^{-1} in August when compared with mid-spring burning only. Late-spring burning combined with atrazine, however, increased indiangrass standing crop 460 kg ha^{-1} in August when compared with late-spring burning only. Based on these results, atrazine provides no benefit over prescribed burning for managing late seral tallgrass prairies, so spring burning date can be altered to meet different management objectives.

SUMMARY

This chapter has provided a brief overview of establishing and managing warm-season grasses. Establishing warm-season grasses requires preparation, proper selection of genetic material, and patience. Following seeding, the primary concern will be managing competition and interference from weeds. Application of herbicides, and in some cases mowing, will generally provide weed control. Once established, prescribed burning can provide many benefits for maintaining a productive stand. Although not exhaustive, this chapter has discussed some important principles for managing weeds in warm-season grass swards. In addition to the references cited, we have included other literature citations as supplemental material for individuals needing more information.

REFERENCES

Abrams, M.D., A.K. Knapp, and L.C. Hulbert. 1986. A ten-year record of aboveground biomass in a Kansas tallgrass prairie: Effects of fire and topographic position. Am. J. Bot. 73:1509–1515.

Aldous, A.E. 1934. Effect of burning on Kansas bluestem pastures. Kansas Agric. Exp. Stn. Tech. Bull. 38. Kansas Agric. Exp. Stn., Manhattan.

Allen, L.J., L.H. Harbers, R.R. Schalles, C.E. Owensby, and E.F. Smith. 1976. Range burning and fertilizing related to nutritive value of bluestem grass. J. Range Manage. 29:306–308.

Anderson, K.L. 1965. Time of burning as it affects soil moisture in an ordinary upland bluestem prairie in the Flint Hills. J. Range Manage. 18:311–316.

Anderson, K.L., E.F. Smith, and C.E. Owensby. 1970. Burning bluestem range. J. Range Manage. 23:81–92.

Bahler, C.C., K.P. Vogel, and L.E. Moser. 1984. Atrazine tolerance in warm-season grass seedlings. Agron. J. 76:891–895.

Baker, R.L., and J. Powell. 1978. Oklahoma tallgrass prairie responses to atrazine with 2,4,-D and fertilizer. p. 681–683. *In* D.N. Hyder (ed.) Proc. Int. Rangeland Cong., Denver, CO. 14–18 Aug. 1978. Soc. or Range Manage., Denver. CO.

Benning, T.L., and T.B. Bragg. 1993. Response of big bluestem (*Andropogon gerardii* Vitman) to timing of spring burning. Am. Midl. Nat. 130:127–132.

Booth, D.T., and M.R. Haferkamp. 1995. Morphology and seedling establishment. p. 239–290. *In* D.J. Bedunah and R.E. Sosebee (ed.) Physiological ecology and developmental morphology. Soc. for Range Manage., Denver, CO.

Bragg, T.B. 1982. Seasonal variations in fuel and fuel consumption by fires in a bluestem prairie. Ecology 63:7–11.

Bragg, T.B., and L.C. Hulbert. 1976. Woody plant invasion of unburned Kansas bluestem prairie. J. Range Manage. 29:19–24.

Christensen, N.L. 1973. Fire and the nitrogen cycle in California chaparral. Science (Washington, DC) 181:66–68.

Cornelius, D.R. 1950. Seed production of native grasses under cultivation in eastern Kansas. Ecol. Monogr. 20:1–29.

Cuomo, G.J., B.E. Anderson, L.J. Young, and W.W. Wilhelm.1996. Harvest frequency and burning effects on monocultures of 3 warm-season grasses. J. Range Manage. 49:157–162.

Curtis, J.T., and M.L. Partch. 1948. Effects of fire on the competition between bluegrass and certain prairie plants. Am. Midl. Nat. 39:437–443.

Curtis, J.T., and M.L. Partch. 1950. Some factors affecting flower production in *Andropogon gerardi*. Ecology 31:488–489.

Dahl, B.E., and P.F. Cotter. 1984. Management of weeping lovegrass in west Texas. Manage. Note 5. Dep. of Range, Wildlife, and Fish. Manage., Texas Tech Univ., Lubbock.

Ehrenreich, J.H. 1959. Effect of burning and clipping on growth of native prairie in Iowa. J. Range Manage. 12:133–137.

Ehrenreich, J.H., and J.M. Aikman. 1957. Effect of burning on seedstalk production of native prairie grasses. Proc. Iowa Acad. Sci. 64:205–212.

Ehrenreich, J.H., and J.M. Aikman. 1963. An ecological study of the effect of certain management practices on native prairie in Iowa. Ecol. Monogr. 33:113–130.

Engle, D.M., T.G. Bidwell, J.F. Stritzke, and D. Rollins. 1990. Atrazine and burning in tallgrass prairie infested with prairie threeawn. J. Range Manage. 43:424–427.

Engle, D.M., R.L. Mitchell, and R.L. Stevens. 1998. Late growing-season fire effects in mid-successional tallgrass prairies. J. Range Manage. 51:115–121.

Ewing, A.L., and D.M. Engle. 1988. Effects of late summer fire on tallgrass prairie microclimate and community composition. Am. Midl. Nat. 120:212–223.

Fick, W.H., and D.E. Peterson. 1995. Musk thistle identification and control. Kansas State Univ. Coop. Ext. Serv. Publ. L-231. Kansas State Univ., Manhattan.

Gay, C.W., and D.D. Dwyer. 1965. Effects of one year's nitrogen fertilization on native vegetation under clipping and burning. J. Range Manage. 18:273–277.

Gillen, R.L., D. Rollins, and J.F. Stritzke. 1987. Atrazine, spring burning, and nitrogen for improvement of tallgrass prairie. J. Range Manage. 40:444–447.

Graves, J.E., and W.E. McMurphy. 1969. Burning and fertilization for range improvement in central Oklahoma. J. Range Manage. 22:165–168.

Hadley, E.B., and B.J. Kieckhefer. 1963. Productivity of two prairie grasses in relation to fire frequency. Ecology 44:389–395.

Hake, D.R., J. Powell, J.K. McPherson, P.L. Claypool, and G.L. Dunn. 1984. Water stress of tallgrass prairie plants in central Oklahoma. J. Range Manage. 37:147–151.

Hanks, R.J., and K.L. Anderson. 1957. Pasture burning and moisture conservation. J. Soil Water Conserv. 12:228–229.

Heath, M.E., and C.J. Kaiser. 1985. Forages in a changing world. p. 3–11. *In* M.E. Heath et al. (ed.) Forages: The science of grassland agriculture. Iowa State Univ. Press, Ames.

Henderson, R.A., D.L. Lovell, and E.A. Howell. 1983. The flowering responses of seven grasses to seasonal timing of prescribed burns in remnant Wisconsin prairie. p. 7–10. *In* R. Brewer (ed.) Proc. of the 8th North American Prairie Conf., Western Michigan Univ., Kalamazoo, MI. 1–4 Aug. 1982. Western Michigan Univ., Kalamazoo.

Hensel, R.1923. Recent studies on the effect of burning on grassland vegetation. Ecology 4:183–188.

Hobbs, N.T., and D.S. Schimel. 1984. Fire effects on nitrogen mineralization and fixation in mountain shrub and grassland communities. J. Range Manage. 37:402–405.

Hobbs, N.T., D.S. Schimel, C.E. Owensby, and D.S. Ojima. 1991. Fire and grazing in the tallgrass prairie: Contingent effects on nitrogen budgets. Ecology 72:1374–1382.

Houston, W.R. 1977. Species susceptibility to atrazine herbicide on shortgrass range. J. Range Manage. 30:50–52.

Houston, W.R., and D.H. Van Der Sluijs. 1975. S-triazine herbicides combined with nitrogen fertilizer for increasing protein on shortgrass range. J. Range Manage. 28:372–376.

Hulbert, L.C. 1988. Causes of fire effects in Tallgrass Prairie. Ecology 69:46–58.

Humphrey, R.R. 1958. The desert grassland: A history of vegetational change and an analysis of causes. Bot. Rev. 24:193–252.

Kay, B.L. 1971. Atrazine and simazine increase yield and quality of range forage. Weed Sci. 19:370–372.

Knapp, A.K. 1985. Effect of fire and drought on the ecophysiology of *Andropogon gerardii* and *Panicum virgatum* in a tallgrass prairie. Ecology 66:1309–1320.

Knapp, A.K., and F.S. Gilliam. 1985. Response of *Andropogon gerardii* (Poaceae) to fire induced high vs. low irradiance environments in tallgrass prairie: leaf structure and photosynthetic pigments. Am. J. Bot. 72:1668–1671.

Knapp, A.K., and T.R. Seastedt. 1986. Detritus accumulation limits productivity of tallgrass prairie. BioScience 36:662–668.

Kucera, C.L. 1970. Ecological effects of fire on tallgrass prairie. *In* Proc. Symp. Prairie Restoration, 1968. Knox College, Galesburg, IL.

Kucera, C.L., and J.H. Ehrenreich. 1962. Some effects of annual burning on central Missouri prairie. Ecology 43:334–336.

Launchbaugh, J.L. 1966. A stand establishment survey of grass plantings in the Great Plains. Nebraska Agric. Exp. Stn. Great Plains Council Publ. 23.Nebraska Agric. Exp. Stn., Lincoln.

Launchbaugh, J.L., and C.E. Owensby. 1970. Seeding rate and first-year stand relationships for six native grasses. J. Range Manage. 23:414–417.

Marietta, K.L., and C.M. Britton. 1989. Establishment of seven high yielding grasses on the Texas High Plains. J. Range Manage. 42:289–294.

Martin, A.R., R.S. Moomaw, and K.P. Vogel. 1982. Warm-season grass establishment with atrazine. Agron. J. 74:916–920.

Masters, R.A. 1995. Establishment of big bluestem and sand bluestem cultivars with metolachlor and atrazine. Agron. J. 87:592–596.

Masters, R.A., R.B. Mitchell, K.P. Vogel, and S.S. Waller.1993. Influence of improvement practices on big bluestem and indiangrass seed production in tallgrass prairies. J. Range Manage. 46:183–188.

Masters, R.A., S.J. Nissen, R.E. Gaussoin, D.D. Beran, and R.N. Stougaard. 1996. Imidazolinone herbicides improve restoration of Great Plains grasslands. Weed Technol. 10:392–403.

Masters, R.A., and K.P. Vogel.1989. Remnant and restored prairie response to fire, fertilization, and atrazine. p. 135–138. *In* T. Bragg and J. Stubbendieck (ed.) Proc. 11th North American Prairie Conf. Univ. of Nebraska, Lincoln. 7–11 Aug. 1988. Univ. of Nebraska, Lincoln.

Masters, R.A., K.P. Vogel, P.E. Reece, and D. Bauer.1990. Sand bluestem and prairie sandreed establishment. J. Range Manage. 43:540–544.

Masters, R.A., K.P. Vogel, and R.B. Mitchell. 1992. Response of Central Plains tallgrass prairies to fire, fertilizer, and atrazine. J. Range Manage. 45:291–295.

McFarland, J.B., and R.B. Mitchell. 1997. Developmental morphology of weeping lovegrass. p. 117–121. *In* M.J. Williams (ed.) Proc. Am. Forage and Grassl. Council, Fort Worth, TX. 13–15 Apr. 1997. Am. Forage and Grassland Council, Georgetown, TX.

McMurphy, W.E., and K.L. Anderson. 1965. Burning Flint Hills range. J. Range Manage. 18:265–269.

Mitchell, R.B., R.A. Masters, S.S. Waller, K.J. Moore, and L.E. Moser.1994a. Big bluestem production and forage quality responses to burning date and fertilizer in tallgrass prairies. J. Prod. Agric. 7:355–359.

Mitchell, R.B., R.A. Masters, S.S. Waller, K.J. Moore, and L.J. Young.1996. Tallgrass prairie vegetation response to spring burning dates, fertilizer, and atrazine. J. Range Manage. 49:131–136.

Mitchell, R.B., L.E. Moser, B.E. Anderson, and S.S. Waller. 1994b. Switchgrass and big bluestem for grazing and hay. Univ. of Nebraska Coop. Ext. Serv. NebGuide G94-1198. Univ. of Nebraska, Lincoln.

Moomaw, R.S., A.R. Martin, and R.N. Stougaard. 1989. Leafy spurge. Univ. of Nebraska Coop. Ext. Serv. NebGuide G87-834. Univ. of Nebraska, Lincoln.

Morrow, L.A., C.R. Fenster, and M.K. McCarty. 1977. Control of downy brome on Nebraska rangeland. J. Range Manage. 30:293–296.

Mullins, S.J., R.B. Mitchell, J.B. McFarland, and D. Holmes. 1997. Spring burning influence on weeping lovegrass yield and forage quality. p. 193–196. *In* M.J. Williams (ed.) Proc. Am. Forage and Grassl. Council, Fort Worth, TX. 13–15 Apr. 1997. Am. Forage and Grassland Council, Georgetown, TX.

Old, S.M. 1969. Microclimate, fire, and plant production in an Illinois prairie. Ecol. Monogr. 39:355–384.

Olson, W.W. 1985. Large scale harvest of native tallgrass prairie. p. 213–215. *In* G. Clambey and R. Pemble (ed.) Proc. 9th North American Prairie Conf., Tri-College Univ., Moorhead, MN. 1 July–1 Aug. 1984. North Dakota St. Univ., Fargo.

Owensby, C.E., and E.F. Smith. 1979. Fertilizing and burning Flint Hills bluestem. J. Range Manage. 32:254–258.

Petersen, N.J.1983. The effects of fire, litter, and ash on flowering in *Andropogon gerardii*. p. 21–24. *In* R. Brewer (ed.) Proc. 8th North American Prairie Conf., Western Michigan Univ., Kalamazoo. 1–4 Aug. 1982. Western Michigan Univ., Kalamazoo.

Petersen, J.L., R.L. Potter, and D.N. Ueckert. 1983. Evaluation of selected herbicides for manipulating herbaceous rangeland vegetation. Weed Sci. 31:735–739.

Rao, M.R., L.H. Harbers, and E.F. Smith. 1973. Seasonal change in nutritive value of bluestem pastures. J. Range Manage. 26:419–422.

Rehm, G., R. Sorensen, and W. Moline. 1976. Time and rate of fertilizer application for seeded warm-season and bluegrass pastures: I. Yield and botanical composition. Agron. J. 68:759–764.

Rehm, G.W. 1984. Yield and quality of a warm-season grass mixture treated with N, P, and atrazine. Agron. J. 76:731–734.

Rice, E.L., and R.L. Parenti. 1978. Causes of decreases in productivity in undisturbed tallgrass prairie. Am. J. Bot. 65:1091–1097.

Risser, P.G., E.C. Birney, H.D. Blocker, S.W. May, W.J. Parton, and J.A. Wiens. 1981. The true prairie ecosystem. Hutchinson Ross Publ. Co. Stroudsburg, PE.

Robocker, W.C., and B.J. Miller. 1955. Effects of clipping, burning and competition on establishment and survival of some native grasses in Wisconsin. J. Range Manage. 8:117–120.

Samson, J.F., and L.E. Moser. 1982. Sod-seeding perennial grasses into eastern Nebraska pastures. Agron. J. 74:1055–1060.

Sauer, C.O. 1944. A geographic sketch of early man in America. Geog. Rev. 34:529–573.

Scifres, C.J. 1980. Brush management: Principles and practices for Texas and the Southwest. Texas A&M Univ. Press, College Station.

Sharrow, S.H., and H.A. Wright. 1977. Effects of fire, ash, and litter on soil nitrate, temperature, moisture, and tobosagrass production in the rolling plains. J. Range Manage. 30:266–270.

Shiflet, T.N., and H.E. Dietz. 1974. Relationship between precipitation and annual rangeland herbage production in southeastern Kansas. J. Range Manage. 27:272–276.

Smith, D.D. 1992. Tallgrass prairie settlement: prelude to the demise of the tallgrass ecosystem. p. 195–199. In D.D. Smith and C.A. Jacobs (ed.) Proc. of the 12th North American Prairie Conf., Cedar Falls, IA. 5–9 Aug. 1990. Univ. of Northern Iowa, Cedar Falls.

Smith, E.F., V.A. Young, K.L. Anderson, W.S. Ruliffson, and S.N. Rogers. 1960. The digestibility of forage on burned and non-burned bluestem pasture as determined with grazing animals. J. Anim. Sci. 19:388–391.

St. John, T.V., and P.H. Rundel. 1976. The role of fire as a mineralizing agent in a Sierran coniferous forest. Oecologia 25:35–45.

Stanford, G., M.H. Frerre, and D.H. Schwaninger. 1973. Temperature coefficient of soil nitrogen mineralization. Soil Sci. 115:321–323.

Towne, G., and C.E. Owensby. 1984. Long-term effects of annual burning at different dates in ungrazed Kansas tallgrass prairie. J. Range Manage. 37:392–397.

USDA-NRCS. 1994. National natural resources inventory database. Misc. Publ. USDA, Washington, DC.

Vallentine, J.F. 1989. Range development and improvements. 3rd ed. Academic Press, San Diego, CA.

Vogel, K.P. 1987. Seeding rates for establishing big bluestem and switchgrass with preemergence atrazine applications. Agron. J. 79:509–512.

Vogl, R.J. 1965. Effects of spring burning on yields of brush prairie savannah. J. Range Manage. 18:202–205.

Vogl, R.J. 1974. Effects of fire on grassland. p. 139–193. In T.T. Kozlowski and C.E. Ahlgren (ed.) Fire and ecosystems. Academic Press, New York.

Waller, S.S., and D.K. Schmidt. 1983. Improvement of eastern Nebraska tallgrass range using atrazine and glyphosate. J. Range Manage. 36:87–90.

Weaver, J.E. 1954. North American Prairie. Johnsen Publ. Co., Lincoln, NE.

Weaver, J.E., and N.W. Rowland. 1952. Effects of excessive natural mulch on development, yield and structure of native grasslands. Bot. Gaz. 114:1–19.

Weaver, J.E., and G.W. Tomanek. 1951. Ecological studies in a midwestern range: The vegetation and effects of cattle on its composition and distribution. Univ. of Nebraska Conserv. Surv. Div. Bull. 31. Univ. of Nebraska, Lincoln.

Wolf, D.D., J.A. Balasko, and R.E. Ries. 1996. Stand establishment. p. 71–85. In L.E. Moser et al. (ed.) Cool-season forage grasses. ASA, CSSA, and SSSA, Madison, WI.

Woolfolk, J.S., E.F. Smith, R.R. Schalles, B.E. Brent, L.H. Harbers, and C.E. Owensby. 1975. Effects of nitrogen fertilization and late-spring burning of bluestem range on diet and performance of steers. J. Range Manage. 28:190–193.

Wright, H.A. 1974. Range burning. J. Range Manage. 27:5–11.

Wright, H.A., and A.W. Bailey. 1982. Fire ecology: United States and southern Canada. John Wiley & Sons, New York.

Additional Noncited References

Adams, D.E., R.C. Anderson, and S.L. Collins. 1982. Differential response of woody and herbaceous species to summer and winter burning in an Oklahoma grassland. Southwest Nat. 27:55–61.

Baldwin, D.M., N.W. Hawkinson, and E.W. Anderson. 1974. High rate fertilization of native rangeland in Oregon. J. Range Manage. 27:214–216.

Biswell, H.H. 1958. Prescribed burning in Georgia and California compared. J. Range Manage. 11:293–297.

Black, A.L. 1968. Nitrogen and phosphorus fertilization for production of crested wheatgrass and native grass in northeastern Montana. Agron. J. 60:213–216.

Cosper, H.R., J.R. Thomas, and A.Y. Alsayegh. 1967. Fertilization and its effect on range improvement in the northern Great Plains. J. Range Manage. 20:216–222.

Gillen, R.L., and W.A. Berg. 1998. Nitrogen fertilization of a native grass planting in western Oklahoma. J. Range Manage. 51:436–441.

Hintz, R.L., K.R. Harmoney, K.J. Moore, J.R. George, and E.C. Brummer. 1998. Establishment of switchgrass and big bluestem in corn with atrazine. Agron. J. 90:591–596.

Kamstra, L.D. 1973. Seasonal changes in quality of some important range grasses. J. Range Manage. 26:289–291.

Kaul, R.B., and S.B. Rolfsmeier. 1987. The characteristics and phytogeographic affinities of the flora of Nine-Mile Prairie, a western tallgrass prairie in Nebraska. Trans. Nebr. Acad. Sci. 15:23–35.

Owensby, C.E., R.M. Hyde, and K.L. Anderson. 1970. Effects of clipping and supplemental nitrogen and water on loamy upland bluestem range. J. Range Manage. 23:341–346.

Patterson, J.K., and V.E. Youngman. 1960. Can fertilizers effectively increase our range land production? J. Range Manage. 13:255–257.

Perry, L.J., Jr., and D.D. Baltensperger. 1979. Leaf and stem yields and forage quality of three N-fertilized warm-season grasses. Agron. J. 71:355–358.

Pyne, S.J. 1982. Fire in America. Princeton Univ. Press, Princeton, NJ.

Rauzi, F., and M.L. Fairbourn. 1983. Effects of annual applications of low N fertilizer rates on a mixed grass prairie. J. Range Manage. 36:359–362.

Rogler, G.A., and R.J. Lorenz. 1957. Nitrogen fertilization of northern Great Plains rangelands. J. Range Manage. 10:156–160.

Samuel, M.J., and R.J. Hart. 1998. Nitrogen fertilization, botanical composition and biomass production on mixed-grass rangeland. J. Range Manage. 51:408–416.

Sauer, C.O. 1950. Grassland climax, fire, and man. J. Range Manage. 3:16–21.

Schultz, R.D., and J. Stubbendieck. 1983. Herbage quality of fertilized cool-season grass–legume mixtures in western Nebraska. J. Range Manage. 36:571–575.

Stewart, O.C. 1951. Burning and natural vegetation in the United States. Geog. Rev. 41:317–320.

Stinson, K.J., and H.A. Wright. 1969. Temperatures of headfires in the southern mixed prairie of Texas. J. Range Manage. 22:169–174.

Vogl, R.J. 1971. The future of our forests. Ecol. Today. 1:6–9.

Warnes, D.D., and L.C. Newell. 1969. Establishment and yield responses of warm-season grass strains to fertilization. J. Range Manage. 22:235–240.

Weaver, J.E. 1965. Native vegetation of Nebraska. Univ. of Nebraska Press, Lincoln.

12 Fertilization of Native Warm-Season Grasses

John J. Brejda

USDA-ARS
Lincoln, Nebraska

Native warm-season (C_4) grasses grow well in acid soils (Jung et al., 1988; Staley et al., 1991), are more efficient in the use of water (Stout et al., 1986), and P (Wuenscher & Gerloff, 1971; Morris et al., 1982), and maintain growth at higher temperatures (Black, 1971) than cool-season (C_3) grasses. These characteristics, combined with their wide range of adaptation and ability to be productive during hot summer months when cool-season grasses are relatively unproductive, has increased the use of native warm-season grasses for pasture and hay in the central and the eastern USA during the past 20 yr (Moser & Vogel, 1995). Native warm-season grasses also are widely grown for erosion control and wildlife habitat (Clubine, 1986). They have been planted on several million hectares of marginal and erosive cropland, roadsides, waterways, railroad and other right-of-ways, and reclamation sites. They comprise the dominant plant species on extensive areas of rangeland within the tallgrass prairie region.

Proper fertilization is an important management tool for improving stand establishment and increasing forage production and quality of native warm-season grasses. Fertilization at the wrong time or with the wrong nutrients is economically inefficient, can lead to stand degradation through invasion of undesirable plant species, and may cause environmental degradation through contamination of surface and groundwater. The purpose of this chapter is to review the response of native warm-season grasses to fertilization and present some general management guidelines for efficient fertilization of these grasses.

NITROGEN-USE EFFICIENCY

High nitrogen-use efficiency (NUE) is agronomically important for effective use of fertilizer N, and for continued growth when soil N supply is low. Nitrogen-use efficiency by a crop is the product of the physiological efficiency (PE) and apparent fertilizer N recovery (AFNR) by the crop (Craswell & Godwin, 1984). Increases in either or both PE and AFNR will increase NUE, and economic return per unit of N applied, and reduce movement of N into surface or groundwater.

Copyright © 2000. Crop Science Society of America and American Society of Agronomy, 677 S. Segoe Rd., Madison, WI 53711, USA. *Native Warm-Season Grasses: Research Trends and Issues.* CSSA Special Publication no. 30.

Table 12–1. Maximum dry matter yield increase, physiological efficiency, apparent fertilizer N recovery, and N-use efficiency of N-fertilized native warm-season grasses.

Cultivar	Location	Range of fertilizer N inputs	Year	Maximum DM yield increase over 0-N control	Physiological efficiency†	AFNR‡	NUE§	Reference¶
		kg ha^{-1}		Mg ha^{-1}	kg kg^{-1}	%	kg kg^{-1}	
Big bluestem								
Pawnee	NE	45–90	1973	6.0	141–145	34–47	50–66	9
			1974	2.7	172–208	14–18	30–31	
Kaw	OK	45–180	1966–1969	4.9	55–109	23–63	13–63	7
Champ	PA	75	1981–1984	0.7–1.6	57	28	16	5
Kaw				0.5–2.6	70	25	17	
Niagara				2.2–3.5	76	49	37	
Pawnee				0.8–2.6	74	31	23	
Switchgrass								
Pathfinder	NE	45–90	1973	2.7	117–163	26–37	31–61	9
			1974	0.9	163–170	1–6	2–10	
Caddo	OK	45–180	1966–1969	7.0	69–157	37–69	25–87	7
Alamo	PA	75	1981–1983	1.3–3.2	--	--	31	5
Blackwell			1981–1984	0.3–4.9	83	37	31	
Caddo				0–3.2	53	41	21	
Kentucky 1625				1.2–4.6	67	50	33	
New Jersey 50				3.1–6.2	100	60	60	
Pathfinder				0.4–3.0	50	35	17	
New Jersey 50	PA	90–180	83–85	3.3–4.9	57–80	44–65	27–40	13
Indiangrass								
Rumsey	MO	56–186	1985	1.8	29–40	27–37	11	1
			1986	2.3	36–51	38–51	14–26	
			1987	2.3	38–47	36–48	14–23	
			1988	4.2	61–83	41–50	25–40	
			1989	4.5	76–108	35–40	27–42	
		78–235	1990	4.3	49–60	37–53	18–31	
Oto	NE	45–90	1973	3.9	147–189	22–29	42–43	9
			1974	0.6	105–112	3–6	3–7	
--	OK	45–180	1966–1969	4.9	69–125	33–52	23–56	7
Cheyenne	PA	75	1981–1982	0.2–1.8	--	--	13	5
Kentucky 591			1981–1984	0.6–2.0	70	29	20	
Nebraska 54				1.2–2.6	86	28	24	
Osage				0.5–2.8	--	--	22	
Western type				0–1.7	51	21	11	

(continued on next page)

In warm-season grasses, PE can be viewed as the efficiency with which they use absorbed N in the synthesis of new leaf and stem tissue. In general, warm-season grasses have greater PE than cool-season grasses. The growth rate of warm-season grasses is higher than that of cool-season grasses across a range of plant N concentrations (Wilson, 1975; Wilson & Brown, 1983; Brown, 1985), they achieve optimum growth rate at lower plant N concentrations, and they are able to maintain growth and leaf development at lower leaf N concentrations than cool-season grasses (Brown, 1978; Wilson & Brown, 1983). In a summary of 119 site-year-species-N rate combinations, PE by native warm-season grasses averaged 80 kg DM

Table 12–1. Continued.

Cultivar	Location	Range of fertilizer N inputs	Year	Maximum DM yield increase over 0-N control	Physiological efficiency†	AFNR‡	NUE§	Reference¶
		kg ha⁻¹		Mg ha⁻¹	kg kg⁻¹	%	kg kg⁻¹	
Little bluestem								
Aldous	PA	75	1981–1984	0.5–2.4	69	21	15	5
Eastern gamagrass								
PMK-24	MO	56–224	1991	0–3.5	0–28	10–88	0–14	2
			1992	3.3–5.5	32–71	33–86	15–44	
			1993	5.5–8.6	--	--	25–52	
Mixture								
	NE	45–90	1967	1.5	104–135	9–13	8–17	10
			1968	0.4	0–47	3–11	0–4	
			1969	4.4	--	--	21–48	
			1970	2.6	83–119	17–29	17–29	
	NE	45–180	1971–1974	1.7–5.1	--	--	10–51	11
	NE	34–134	1974–1979	2.4–4.8	--	--	12–75	12
Native range								
	KS	37–75	1963	1.7	81–97	16–27	13–23	8
	OK	56–112	1965	0.1–2.1	--	--	0–28	3
	OK	45–90	1966	3.2	--	--	36–44	4
			1967	4.2	--	--	46–50	
Native meadow								
	OK	45	1965	1.1	48–105	21–23	10–24	6
			1966	1.5	96–137	20–24	19–33	
			1967	1.7	99–113	28–33	28–38	
			1968	2.1	127–133	19–38	26–48	
			1969	2.4	--	--	20–54	

† Calculated as [(forage yield$_F$ − forage yield$_c$)/(N uptake$_F$ − N uptake$_c$)] where F = fertilized plants and C = unfertilized control.
‡ Apparent fertilizer N recovery = [(N uptake$_F$ − N uptake$_c$)/fertilizer N applied] × 100 where F = fertilized plants and C = unfertilized control.
§ Nitrogen-use efficiency = [(forage yield$_F$ − forage yield$_c$)/(fertilizer N applied)] where F = nitrogen fertilized plants and C = unfertilized control.
¶ References: (1) Brejda et al., 1995; (2) Brejda et al., 1996; (3) Gay and Dwyer, 1965; (4) Graves and McMurphy, 1969; (5) Jung et al., 1990; (6) McMurphy, 1970; (7) McMurphy et al., 1975; (8) Moser and Anderson, 1964; (9) Perry and Baltensperger, 1979; (10) Rehm et al., 1972; (11) Rehm et al., 1976; (12) Rehm, 1984; (13) Staley et al., 1991.

kg⁻¹ N (SD = 41) across a range of N rates (Table 12–1). In general, PE declined with increasing rates of N; however, regression of PE on N rate accounted for only 21% of the variation in PE, suggesting that a multitude of environmental and management variables also affect PE. Improvement in PE in warm-season grasses will require a better understanding and control of environmental and management factors that effect PE.

Apparent fertilizer N recovery reflects the efficiency of the crop in obtaining fertilizer N from the soil. In annual grain crops, AFNR in above ground plant parts averages about 50% (Craswell & Godwin, 1984). In contrast, a summary of

131 site-year-species-N rate combinations showed that AFNR by native warm-season grasses averaged only 37% (SD = 20%) (Table 12–1). The greater AFNR of annual grain crops could be attributed to the fact that they are usually planted on better agricultural land, and the timing and placement of fertilizer N is made to maximize N uptake. Despite this management difference, the low average AFNR by native warm-season grasses indicates there is considerable need for improvement in AFNR with these species.

In an summary of 215 site-year-species-N rate combinations, NUE in native warm-season grasses averaged 29 kg DM kg^{-1} N ha^{-1} (SD = 15), and in general, declined with increasing rates of N (Table 12–1). In comparison, in a North Dakota study, the 9-yr average NUE for seven cool-season grasses fertilized with 45 kg N ha^{-1} was 38 kg DM kg^{-1} N ha^{-1}, and 14 kg DM kg^{-1} N ha^{-1} when fertilized with 225 kg N ha^{-1} (Power, 1985). Based on this comparison, it would appear that NUE by warm- and cool-season grasses are similar; however, NUE in the 215 site-year-species combinations ranged as high as 87 kg DM kg^{-1} N ha^{-1} (Table 12–1) indicating that under certain conditions, NUE by native warm-season grasses can be very high. The challenge agronomists and range scientists face is to understand and control the multitude of environmental and management factors that influence PE, AFNR, and ultimately, NUE in native warm-season grasses, to consistently achieve the most efficient use of N fertilizer.

ENVIRONMENTAL EFFECTS ON WARM-SEASON GRASS RESPONSES TO NITROGEN FERTILIZATION

Climate

Environmental conditions regulate N transformations in the soil, affecting the amount of soil and fertilizer N available to the grasses, and the efficiency with which the grasses use absorbed N to produce dry matter (Craswell & Godwin, 1984). Variability in weather makes N management difficult because the flux of N through every pathway in the soil is either directly or indirectly influenced by temperature or moisture. These same environmental variables also strongly influence warm-season grass growth and phenological development. Soil temperature in the spring controls when warm-season grasses initiate growth. The accumulation of heat units influences the growth rate of warm-season grasses during the growing season, and ultimately, crop sink strength for available N. Warm soil temperatures also increase N mineralization from soil organic matter and N losses through NH_3 volatilization and denitrification (Craswell & Godwin, 1984).

Annual precipitation has been shown to be the most important factor in explaining primary production in grasslands in the central region of the USA (Sala et al., 1988). The amount and distribution of rainfall also affects water flow through the profile, soil water content, and soil aeration. These factors influence the rate of N loss from the available N pool through NH_3 volatilization, denitrification, and nitrate leaching. Precipitation carries N onto the land via atmospheric deposition, but it also carries N off the soil surface or below the root zone. Ammonium ions

(NH_4^+) are less mobile than NO_3^- and less likely to be leached from the root zone; however, in well aerated soils with adequate moisture and moderate temperature, NH_4^+ and urea can be rapidly converted to NO_3^- by soil microorganisms (Bohn et al., 1979; Killham, 1994). In contrast, loss of NO_3^-–N through denitrification, in which NO_3^- is converted to N_2O or N_2, occurs when the soil is at or near water saturation and O_2 diffusion into the soil is low (Killham, 1994). The main climatic variable influencing denitrification is rainfall, which influences the water content of the soil and thus soil aeration (Craswell & Godwin, 1984).

Because of wide year-to-year and within year variation in environmental conditions, warm-season grass responses to N fertilization can vary significantly between environments. In southeast Missouri, Brejda et al. (1995) fertilized indiangrass [*Sorghastrum nutans* (L.) Nash] with up to 168 kg N ha^{-1} for five consecutive years from 1985 through 1989. They reported significant year-to-year variation in indiangrass yield responses to N fertilization, ranging from a low of 1.8 Mg ha^{-1} in 1985 to >4.5 Mg ha^{-1} in 1989, at the 168 kg N ha^{-1} rate (Table 12–1). During that same time period, AFNR by indiangrass ranged from a low of 27% in 1985 to >50% in 1986, 1988, and 1990 (Table 12–1). In north central Missouri, Brejda et al. (1996) fertilized eastern gamagrass [*Tripsacum dactyloides* (L.) L.] with a single spring N application of up to 224 kg N ha^{-1} for three consecutive years. Eastern gamagrass yield increases over the no N control treatment also varied significantly from year-to-year, peaking at 3.5 Mg ha^{-1} in 1991, 5.5 Mg ha^{-1} in 1992, and 8.6 Mg ha^{-1} in 1993, at the 224 kg N ha^{-1} rate. In a 3-yr study with big bluestem (*Andropogon gerardii* Vitman), switchgrass (*Panicum virgatum* L.), and indiangrass in Iowa, Hall et al. (1982) reported that maximum yield responses to N ranged from 0 to 45% in 1974, 10 to 45% in 1975, and 70 to 105% in 1976. Significant year-to-year variation in warm-season grass response to N fertilization also has been observed with several big bluestem, switchgrass, and indiangrass cultivars in Pennsylvania and warm-season grass mixtures in Nebraska (Table 12–1). In these studies, large differences in growing season precipitation amount and distribution between years were important factors contributing to the wide variation in warm-season grass yield responses to N fertilization.

Wide year-to-year variation in warm-season grass yield responses to N fertilizer typically produce significant year × N rate interactions. Yield responses to increasing rates of N may be linear one year, suggesting the grasses may be able to respond to even greater rates of N, and curvilinear the next. The optimum N-rate for warm-season grass yields showing a curvilinear response to increasing rates of N will also vary from year-to-year. Such patterns make management recommendations difficult, if they are based solely on yield response curves to N fertilizer.

Soil

Many of the effects soils have on N availability and warm-season grass responses to N fertilization result from interactions between the soil and its local environment. In an examination of primary production within the central grassland region of the USA, Sala et al. (1988) concluded that at a given site, annual precipitation and soil water holding capacity had the strongest influence on primary production. When precipitation was <370 mm yr^{-1}, sandy soils with low water-holding capacity were more productive than loamy and fine-textured soils with high water-holding capacity, while the opposite pattern occurred when precipitation was >370 mm yr^{-1}.

Some soil properties can directly influence N availability and warm-season grass responses to N fertilizer. Mineralogy, texture, organic matter content, water-holding capacity, and pH directly or indirectly influence plant growth and N cycling. Soil mineralogy influences NH_4^+ fixation, in which NH_4^+ ions become bound between the 2:1 layer silicate lattices of clay particles. Ammonium fixation can be a problem in soils high in vermiculite and weathered micas. Ammonium fixation generally decreases with soil acidification and increases with soil liming (Bohn et al., 1979).

Soil texture influences soil porosity, which in turn influences aeration, water infiltration rates, water-holding capacity, and the rate at which N and other nutrients leach through the soil. Soil texture also influences microbial activity that controls the cycling of N, P, S, and C through immobilization and mineralization processes. In general, coarser texture soils have higher microbial activity and thus more rapid nutrient cycling than finer texture soils (Elliott et al., 1980; Coleman et al., 1983).

Soil organic matter content also influences water-holding capacity, but more importantly, organic matter contains the largest N pool in the soil. Nitrogen mineralized from soil organic matter often comprises the major proportion of N available to the plant each year (Craswell & Godwin, 1984). Assessment of N mineralization rates from the soil organic matter pool is essential for determining the N-supplying capacity of the soil and N fertilizer needs of the crop to ensure that fertilizer N is used efficiently.

In Pennsylvania, Stout and Jung (1995) grew switchgrass on four soils with different N supply capacities. They reported that biomass and N accumulation were greatest on the soil with the highest native N level; however, the yield response of switchgrass to increasing rates of N and the AFNR were lowest on this same soil. Soil water-holding capacity also can affect N uptake. In a study of switchgrass response to increasing rates of N on three soils with different water-holding capacities, Staley et al. (1991) reported that soil N uptake increased as soil water-holding capacity increased.

MANAGEMENT EFFECTS ON WARM-SEASON GRASS RESPONSES TO NITROGEN FERTILIZATION

Grass Species and Cultivars

Warm-season grass responses to N fertilization are affected by the species and cultivar, N source, N timing, burning, haying, and grazing. The data in Table 12–1 suggests there is some variation in PE, AFNR, and NUE between different warm-season grass species and cultivars; however, variation between studies and year within studies makes drawing conclusions difficult.

The data in Table 12–1 suggests that in general, yield responses to N fertilization of native prairies is lower than for seeded warm-season grass stands. There are several possible explanations for this. Native prairies are comprised of a large number of species, some of which respond poorly to increasing rates of N. In addition, warm-season grasses present in native prairies have not been selected and bred for higher yields as have cultivars used in seeded warm-season grass stands. Most native prairie sites within the tallgrass prairie region have never been cultivated because of soil limitations at these sites. The soil factors that discouraged cultivation also may limit the response of the prairie vegetation to N fertilization. Finally, Brejda (1996) observed that arbuscular mycorrhizal fungi AMF from native prairies was less effective in nutrient uptake and promoting warm-season grass growth than AMF from seeded warm-season grass stands.

Nitrogen Source

Numerous studies have reported urea to be less efficient than ammonium nitrate or ammonium sulfate as a N source for perennial grasses. Ammonia volatilization losses from N applied as urea can be appreciable. These losses are strongly influenced by soil and climatic conditions. Alkaline soil conditions (pH > 7.0) and the presence of plant residues on the soil surface increase NH_3 volatilization losses (Hargrove, 1988). On neutral and moderately acid soils, however, urea may be an equally efficient source of N for warm-season grasses. Efficiency may be improved if the stands are burned to remove plant residues in the spring prior to fertilizer application.

On an acid soil (pH 5.57) in southwest Missouri, indiangrass and caucasian bluestem [*Bothriochloa caucasia* (Trin.) C.E. Hubbard] stands were burned every spring and fertilized with increasing rates of urea or ammonium-nitrate (NH_4NO_3) (Brejda et al., 1995). Forage yield, crude protein concentration, or both were greater with NH_4NO_3 compared with urea in 3 out of 6 yr with indiangrass, and 4 out of 6 yr with caucasian bluestem. There were no differences between the two fertilizer sources in the other years. In Oklahoma, Westerman et al. (1983) applied five N sources to bermudagrass (*Cynodon dactylon* L.) on five different soils ranging in pH from 4.7 to 6.9. They reported that the relative efficiency of the different N sources, averaged across sites, was ammonium sulfate > urea-ammonium-nitrate > urea > anhydrous ammonia.

Nitrogen Timing

Proper timing of N applications is important for both efficient uptake and use of fertilizer N, and to prevent stand degradation by invasion with cool-season grasses. In Nebraska, Rehm et al. (1976, 1977) reported that early spring application of N on mixed warm-season grasses encouraged invasion by intermediate wheatgrass [*Agropyron intermedium* (Host) Beauv.]. With a late spring application, forage crude protein concentrations were greater and cool-season grass invasion did not occur, provided the N rate was <90 kg ha^{-1}.

Late summer N applications also can encourage cool-season grass invasion. On an upland bluestem range site in Kansas, application of 56 kg N ha^{-1} on 1 July stimulated Kentucky bluegrass (*Poa pratensis* L.) invasion (Owensby et al., 1970). The timing of the N application was too late in the growing season for efficient uptake and use by the warm-season grasses. This resulted in residual carryover of fertilizer N into the fall when it stimulated Kentucky bluegrass growth.

Burning

Burning increases N cycling rates in the tallgrass prairie by removing surface residues and increasing soil temperature in the spring. This stimulates N mineralization from soil organic matter (Risser & Parton, 1982), making it available for uptake by the grass. In addition, dead warm-season grass foliage and stems are a N sink during the first 2 yr of decomposition (Seastedt, 1988). Combustion of dead litter reduces this N sink and thus reduces N immobilization; however, combustion of litter and herbaceous residues also reduces the amount of N returned to the soil through litterfall, which can result in N losses of up to 20 kg N ha^{-1} (Seastedt, 1988). The quantity of N lost by burning is influenced by the amount of biomass previously removed by grazing or haying. In Kansas, N combustion losses on ungrazed plots were 18 kg ha^{-1} yr^{-1}, but on grazed plots N losses were only 9 kg ha^{-1} yr^{-1} (Hobbs et al., 1991).

When burning is used in combination with N fertilization, it can increase NUE in warm-season grasses. Initiation of new growth in the spring is often earlier with burned warm-season grass stands, and photosynthesis has been shown to be higher for burned relative to unburned plants (Svejcar & Browning, 1988). These changes can increase PE in warm-season grasses. In addition, increased leaf area and tillering has been observed with burned compared with unburned plants. This enhances sink strength and N uptake by burned plants (Svejcar & Browning, 1988), increasing AFNR. Increased PE and AFNR due to burning will increase NUE by burned relative to unburned warm-season grasses. Increases in PE, AFNR, and NUE can vary with the burning frequency. In Kansas, forage production in the Flinthills prairie increased an average of only 9% on unburned sites in response to application of 200 kg N ha^{-1} (Seastedt et al., 1991). In contrast, forage production on infrequently burned sites increased at average of 45%, and forage production on annually burned sites increased an average of 68% in response to application of 200 kg N ha^{-1}. Seastedt et al. (1991) attributed the greater yield response to N fertilizer on annually burned sites to lower N availability.

As with N fertilization, warm-season grass responses to burning is influenced by the environment. In Kansas, annual burning increased net primary production of native prairie on lowland sites where soil moisture availability was greater, but had no effect on net primary production on drier upland sites (Briggs & Knapp, 1995). Decreases in forage production following burning have also been reported. In Nebraska, Cuomo et al. (1996) reported that annually burning big bluestem, switchgrass, and indiangrass stands in which the forage was harvested during the growing season decreased annual forage yields compared with unburned plots. The decrease was greatest with May compared with April or March burns for all three species. In summarizing the effects of burning on warm-season grass yields, Cuomo et al. (1996) concluded that increased forage yields following burning were generally found in high precipitation zones when burning occurred after several years of little or no forage removal. Burning may not increase warm-season grass yields in regions of inadequate or below normal precipitation. These same environmental conditions also limit warm-season grass response to N fertilizer.

Burning can reduce competition for ferilizer N by undesireable cool-season grasses. Owensby and Smith (1979) applied 0, 45, and 90 kg N ha^{-1} in combination with fire to a Flinthills bluestem prairie. Application of 90 kg N ha^{-1} increased Kentucky bluegrass invasion on both burned and unburned plots. The increase in Kentucky bluegrass, however, was small on burned plots, compared with a large increase on unburned plots.

Burning also can result in increased crude protein concentrations in warm-season grasses (Svejcar & Browning, 1988; Cuomo & Anderson, 1996), particularly early in the growing season, which may result from greater soil N mineralization. Increases in crude protein content with burning, however, are usually less than increases observed with N fertilization. Fertilization of big bluestem, switchgrass, and indiangrass with 66 kg N ha^{-1} increased forage crude protein content an average of 18%, from 64 to 76 g kg^{-1} for unburned and burned plots respectively, with 65% of the additional protein in the form of rumen degradable protein (Cuomo & Anderson, 1996). In contrast, burning increased early season (mid-June) crude protein content only 7%, and none of the increase was in the form of rumen degradable protein.

Grazing and Haying

Grazing and haying have a significant influence on warm-season grass response to N fertilization. In a simulation analyses of the N cycle in the tallgrass prairie, grazing increased N cycling rates through higher decomposition rates and increased N mineralization in the plant material consumed by cattle (*Bos taurus*) and returned as urine and feces (Risser & Parton, 1982). Grazing also increased N mineralization from soil organic matter and increased formation of new soil organic N.

Both grazing and haying resulted in an export of N from the stand, but losses were much larger with haying compared with grazing. Risser and Parton (1982) estimated that with season-long grazing at moderate stocking rates (2.4 ha animal unit (AU)$^{-1}$), only 4.4 kg N ha^{-1} yr^{-1} was lost through direct removal by the animal and volatilization from urine and dung patches. In contrast, with haying, in which up to 90% of the above ground biomass is removed from a site, >100 kg N ha^{-1} yr^{-1} may be removed (McMurphy et al., 1975; Staley et al., 1991; Brejda et al., 1995, 1996). Because of greater N removal with haying, higher N rates are needed to maximize forage yields for hay production than with grazing.

A poorly understood effect that grazing or haying can have on warm-season grass response to N fertilizer is through its impact on N cycling through grass roots. In the tallgrass prairie, below ground biomass production comprises 48 to 64% of net primary production (Stanton, 1988). The N content of warm-season grass roots is low. At a tallgrass prairie site in Kansas, root N concentration averaged 4.9 g kg^{-1} (Seastedt, 1988), and in northeastern Oklahoma, root N concentration averaged 5.3 g kg^{-1} during the growing season (Risser & Parton, 1982). The large root mass combined with an average turnover of 30 to 40% yr^{-1} (Dahlman & Kucera, 1965; Hayes

& Seastedt, 1987) can result in appreciable amounts of N mineralized from roots each year. In northeast Oklahoma, Risser and Parton (1982) estimated that 30 kg N ha^{-1} yr^{-1} was released to the available N pool, and in Kansas, Seastedt (1988) calculated a net release of 15 to 20 kg N ha^{-1} yr^{-1} in the absence of haying, grazing, or N fertilizer.

Because of their high C/N ratio, grass roots, soil organic matter, and the microbial pool can immobilize significant amounts of fertilizer N (Oswalt et al., 1959; Power, 1968; Garcia & Rice, 1994). Using simulation studies of the N cycle in the tallgrass prairie, Risser and Parton (1982) estimated that >26% of fertilizer N was immobilized in grass roots, and an additional 9% was immobilized in soil organic matter. Following defoliation, plant root growth stops and a portion of the existing root system may die. Decomposition of fertilizer N enriched warm-season grass roots will result in mineralization of N and an increase in the available N pool (Power, 1968). Garcia and Rice (1994) reported that mineralization of N from microbial biomass was enhanced in response to mowing. They attributed this response to reduced carbon input to the soil through above ground biomass removal and root growth suppression. The decrease in microbial N was accompanied by a transitory increase in inorganic N. These results suggest that N fertilization and haying alters the sink size and turnover rates of N in the soil, resulting in the establishment of new equilibrium levels between rates of immobilization and mineralization in warm-season grass stands.

BIOLOGICAL NITROGEN FIXATION

Free-Living and Associative Nitrogen Fixation

Renewed emphasis on economical, sustainable and environmentally safe N sources has resulted in an increased interest in biologically fixed N for native warm-season grasses. Potential sources of biologically fixed N for warm-season grasses include N-fixation by free-living diazotrophs, associative N-fixation, and symbiotic N-fixation. Cyanobacteria are free-living N-fixing organisms that were found to be locally abundant in a relict tallgrass prairie in Ohio (DuBois & Kapustka, 1983). It was estimated, however, that <10 kg N ha^{-1} yr^{-1} was fixed by these organisms. Although the amount of N contributed annually to native warm-season grasses is low, the cumulative effect over many year may be important in the buildup of organic N content of grassland soils.

Associative N-fixation is defined as N-fixation by nonsymbiotic diazotrophic bacteria under the direct influence of a host plant (Klucas, 1991). Associative N-fixing bacteria can colonize live and dead roots, both externally and internally (Bashan & Levanony, 1990; Sumner, 1990). With native warm-season grasses, associative N-fixation has been reported with switchgrass (Tjepkema, 1975; Tjepkema & Burris, 1976; Morris et al., 1985), and eastern gamagrass (Brejda et al., 1994). However, there is little information on the bacterial species associated with native warm-season grasses. *Azospirillum brasilense* was isolated from the rhizosphere of eastern gamagrass in northern Missouri (Brejda et al., 1994) and *Klebsiella planticola* has been isolated from the rhizosphere of switchgrass in southeastern Nebraska (J.J. Brejda, 1996, unpublished data). In both cases, the bacteria were isolated from surface sterilized roots, suggesting that the bacteria were living within the cortex of the roots.

Rhizosphere conditions favorable for associative N-fixation include low availability of combined N, high soil moisture content, a readily available energy source, and low O_2 concentrations (Klucas, 1991). The rhizosphere conditions that favor associative N-fixation, however, also limit its potential benefits. The high soil moisture requirement limits associative N-fixation to moist sites, and the low availability of combined N required for N-fixation will limit forage yields. Further, the quantities of N-fixed by associative N fixation have been highly variable and smaller than expected, often <5 kg ha^{-1} yr^{-1} (Klucas, 1991; Giller & Wilson, 1991). In addition, associative N-fixing bacteria reduce atmospheric N for their own growth. This N does not become available to the plant until the bacteria die and the microbial N is mineralized. Under these conditions, the plant must compete with all other rhizosphere organisms for the microbial N, making the transfer of N between the bacteria and the plant inefficient (Sloger & van Berkum, 1991).

Despite these limitations, enhanced plant growth has been reported following inoculation of plants with associative N-fixing bacteria. Because only small amounts of reduced N are generally believed to be provided to the plant on an annual basis, increases in plant growth following inoculation have been attributed to other plant growth promoting effects of the bacteria. These include production of plant growth promoting hormones, increased mineral and water uptake, increased root surface area and diameter, and density and length of root hairs (Okon 1985; Bashan & Levanony, 1990; Sumner, 1990). Thus, associative N-fixing bacteria still hold promise as a biological means of enhancing plant growth in native warm-season grasses, but it does not appear to be through N-fixation.

Symbiotic Nitrogen Fixation

The potential for providing N to warm-season grasses through symbiotic N-fixation by legumes will vary between seeded stands and native prairies, seral stages within a prairie, and with legume density and plant age. Native legume species vary widely in nodular number, nodular weight per plant, nodule efficiency, and total N fixation. In an evaluation of five native legume species, the rank of species, from greatest to lowest, in total N-fixation was partridge pea (*Cassia chamaecrista* L.), roundhead lespedeza (*Lespedeza capitata* Michx.), sensitive brier [*Schrankia nuttalli* (DC.) Standl.], leadplant [*Amorpha canescens* (Nutt.) Pursh], and silverleaf scurfpea (*Psoralea argophylla* Pursh) (Becker & Crockett, 1976). The authors concluded that the two species with the greatest total N-fixation, partridge pea and roundhead lespedeza, tend to be early seral species, whereas the two species with lowest total N-fixation, leadplant and scurfpea, are climax or late seral species. This suggests that in climax grassland communities, lower quantities of N may be fixed because of the species composition of the legumes in the community.

In Oklahoma, Kapustka and Rice (1978) used the data from Becker and Crockett (1976) and the density of common native legumes to estimate total quantities of N fixed during a growing season. They concluded that <1 kg N ha^{-1} yr^{-1} was fixed at their study site. Similar low values have been estimated for native legumes in a Canadian grassland site (Vlassak et al., 1973).

Kneebone (1959) studied 19 native legume species for use in reseeding western Oklahoma rangelands. He rated 10 of the 19 species as fair or poor in nodule formation. In addition, Kneebone (1959) examined legumes in native stands and reported that nodules were difficult to find on old plants. The evidence suggests that native legumes are a poor source of N for grasses in native prairies.

In contrast to native prairies, legumes interseeded into seeded warm-season grass stands have been shown to increase both forage yield and quality. In central Kansas, Posler et al. (1993) examined binary mixtures of 'Blackwell' switchgrass, 'Osage' indiangrass, and 'El Reno' sideoats grama (*Bouteloua curtipendula* Michx.) in combination with five native and one introduced legume species. Forage yields of all binary mixtures, except the switchgrass–leadplant mixture, were significantly greater than grass only. In some cases this resulted from additional forage production by the legumes; however, with purple prairieclover (*Dalea purpurea* Vent.) and indiangrass, and with sensitive brier and sideoats grama, grass yields were greater when grown in combination with the legume, than when grown alone, suggesting that these two legumes were supplying fixed N to the grasses (Posler et al., 1993). Inclusion of legumes significantly improved crude protein concentration, but did not consistently improve forage digestibility over that of grasses alone. In general, the four native legumes persisted well in combination with all three warm-season grasses. The introduced cool-season legume cicer milkvetch (*Astragalus cicer* L.), however, was not compatible because it developed a thick, dense canopy prior to initiation of growth of the warm-season grasses.

In two seeding trials in central Iowa, George et al. (1995) evaluated forage yields of established stands of 'Cave-in-rock' switchgrass interseeded with 10 cool-season legumes, compared with fertilization with 0, 60, 120, and 240 kg N ha^{-1}. In the establishment year, interseeding legumes did not affect June forage yields, but produced 9% greater yields than nonfertilized switchgrass stands in July. In both trials, second year total forage yields of switchgrass stands were greater when interseeded with 'Norcen' and 'Fergus' birdsfoot trefoil (*Lotus corniculatus* L.), or 'Mammoth' red clover (*Trifolium pratense* L.) than switchgrass stands fertilized with 240 kg N ha^{-1}.

Switchgrass stem density during the establishment year was not affected by interseeding legumes; however, during the second year, switchgrass stem density was lower in legume interseeded stands than in stands treated with no N (Blanchet et al., 1995). The reduction in switchgrass stem density was greatest when interseeded with both birdsfoot trefoil cultivars and a Norcen trefoil–Redland II red clover mixture, and intermediate when interseeded with 'Alfagraze' alfalfa (*Medicago sativa* L.) or both red clover cultivars. This suggests that several of the legumes had replaced a portion of the switchgrass stand.

NITROGEN RECOMMENDATIONS FOR WARM-SEASON GRASSES

Nitrogen requirements for native warm-season grasses are generally less than for cool-season grasses. Ideally, the N application rate should be specific for the grass species, the yield potential of the stand for the given soil and climate, whether the stand will be used for grazing or hay production, and the economic return per unit of N applied. Of these factors, the economic return from N fertilization should be considered first, because in some situations it may be cheaper to purchase additional forage or rent pasture than to fertilize.

Assuming it is economical to apply N fertilizer, a practical starting point is to replace the amount of N removed in the harvested forage. In general, about 13 kg N will be removed in 1000 kg of hay with a crude protein content of 80 g kg^{-1} of forage. The actual protein content of warm-season grass hay will vary with species, cultivar, and stage of maturity at harvest.

The yield potential of warm-season grasses will vary with the N-supplying capacity of the soil and growing season precipitation amounts and distribution. In Nebraska, N recommendations for pasture and hayland vary with climatic zones within the state. For southeast Nebraska, where precipitation amounts are the highest, application of 67 to 100 kg N ha^{-1} yr^{-1} is recommended for pasture and 85 to 110 kg N ha^{-1} yr^{-1} is recommended for hayland. Recommended N rates decrease from east to west across the state, as growing season precipitation amounts decrease. Iowa recommends application of 90 to 110 kg N ha^{-1} yr^{-1} for seeded warm-season grasses. In Kansas and Missouri, the current N recommendation for seeded warm-season grasses is 67 kg N ha^{-1} yr^{-1}, irrespective of soil type and climatic zone. The Kansas recommendation for native prairies, however, is only 33 kg N ha^{-1} yr^{-1}, because higher N rates promote invasion by cool-season grasses. All four states recommend that N application be delayed until after green-up in the spring to reduce stimulation of cool-season grasses and minimize N losses by leaching, runoff, volatilization, or denitrification.

In general, N fertilizer is not used when seeding new warm-season grass stands unless a herbicide also is used because undesirable species may be stimulated more by N application than the seeded species. Once a warm-season grass stand is established, however, N fertilization has been shown to increase the density of seeded warm-season grass stands, especially on marginal and eroded cropland where native soil N levels are low (Warnes & Newell, 1969). However, N fertilization will not overcome improper management or overgrazing of native warm-season grasses. Appropriate management is as necessary on seeded warm-season grass stands and prairies as on cropland. Warm-season grass stands should be evaluated before N fertilization is considered to ensure they are being managed properly in terms of intensity and season of use. Otherwise, N applications may not only be uneconomical but also lead to further stand degradation.

Species composition of the stand and potential for undesirable shifts in the vegetation or invasion by weeds also will affect the appropriate N application rates. In Kansas, the recommended N application rate for native prairie is only about one-half that recommended for seeded stands to prevent shifts in species composition. The N rate can be increased if other management practices are included that help maintain dominance by the warm-season grasses. With seeded warm-season grass stands these may include periodic herbicide application, fire, and appropriate timing, duration, and intensity of grazing. On native range, fire and proper grazing management may be the only management option available because treatment with a herbicide could injure or kill native cool-season grasses and forbs, which may be important forage species.

Information on appropriate N rates and timing are needed for producers interested in using highly competitive legumes with native warm-season grasses; however, currently there is no information on N requirements of native warm-season grass–legume mixtures. Research with cool-season grass–legume mixtures has shown that grasses respond to N fertilization much more than do legumes. As a result, N fertilization can result in elimination of legumes from the stand; however, many native warm-season grasses are not very competitive with cool-season legumes when subjected to frequent defoliation. Therefore, application of N fertilizer may help maintain a warm-season grass component in the mixture when interseeded with more competitive cool-season legumes. Care must be taken, however, to prevent invasion by cool-season grasses.

PHOSPHORUS, POTASSIUM, AND LIME

Phosphorus

Native warm-season grasses have lower tissue P requirements and greater P-use efficiency than cool-season grasses (Morris et al., 1982) and appear to be able to meet their P requirements on soils too low in P to support cool-season grasses. In a greenhouse pot study using a low P soil containing 2 to 5 mg P kg^{-1}, Wuenscher and Gerloff (1971) compared growth, P content, and response to P fertilizer by little bluestem [*Schizachyrium scoparium* (Michx.) Nash] and Kentucky bluegrass. Little bluestem absorbed more soil P, used absorbed P more efficiently, and produced greater yields than Kentucky bluegrass. In a field study in Pennsylvania using a low P soil (5 mg kg^{-1} Bray and Kurtz no. 1), Morris et al. (1982) compared warm- and cool-season grasses treated with 0 or 448 kg P ha^{-1}. Tissue P concentration of the warm-season grasses (0.7 to 1.1 g P kg^{-1}) was about one-half that of the cool-season grasses (1.4 to 2.2 g P kg^{-1}), but warm-season grass yields were up to three times greater than cool-season grass yields on both the high and low P soils. Total P uptake was similar between the warm- and cool-season grasses, but the warm-season grasses had greater P-use efficiency.

The ability of native warm-season grasses to meet their P requirements in soils too low to adequately meet the P requirements of cool-season grasses is partially a result of the symbiotic relationship they form with arbuscular mycorrhizal fungi (Hetrick et al., 1988; 1991; Brejda et al., 1993). Arbuscular mycorrhizal fungi also colonize the roots of cool-season grasses, but cool-season grasses are believed to be less dependent upon mycorrhiza for uptake of P and other nutrients because of their finer root architecture (Hetrick et al., 1991).

In several fertilizer studies in Nebraska, Kansas, and Oklahoma, warm-season grass yields were depressed when P was applied alone (0 N treatment)(Table 12–2). In most studies, the yield depression was <250 kg ha^{-1} (Table 12–2). In a study of N and P cycling within a tallgrass prairie in Kansas, Seastedt (1988) observed that N was immobilized in root detritus during the first 2 yr of decomposition, but P was readily mineralized. Application of P fertilizer, without additional N, may stimulate microbial growth and N immobilization, reducing N availability to the warm-season grasses, and causing a yield depression at 0-N rates.

Phosphorus fertilization often produce little or no response in native warm-season grasses when applied in combination with low rates of N. In Oklahoma, increasing rates of P (0, 50, 100, and 200 kg P ha^{-1}) had no effect on growth and tissue P content of little bluestem fertilized with 100 kg N ha^{-1} (Sharpley & Reed, 1982). In Pennsylvania, yields of big bluestem and indiangrass fertilized with 56 kg N ha^{-1} showed no response to application of 448 kg P ha^{-1} on a low P soil (Morris et al., 1982). In the same study, switchgrass yields increased only 12% and little bluestem yields increased only 15% with application of 448 kg P ha^{-1}. In contrast, cool-season grass yields increased an average 35% with application of 448 kg P ha^{-1}.

As N rate increases, yield and P-use efficiency by native warm-season grasses also increases (Table 12–2), producing a significant N × P interaction in several studies (McMurphy et al., 1975; Taliaferro et al., 1975; Rehm, 1984; 1990). At N rates of 90 kg N ha^{-1} or greater, P-use efficiency by warm-season grasses may exceed N-use efficiency by as much as two-fold, producing >100 kg dry matter kg^{-1} P ha^{-1} (Table 12–2). Phosphorus-use efficiency decreases, however, as the P rate increases, and most fertilizer studies have evaluated only low rates of P, typically in the range of 45 kg P ha^{-1} or less.

The effect of P fertilization on warm-season grass quality is small and inconsistent. In Nebraska, P fertilization of a seeded warm-season grass mixture increased forage crude protein content when applied in combination with N in one out of 3 yr (Rehm et al., 1977). In contrast, several studies in Nebraska and Oklahoma reported that P fertilization decreased crude protein content in both native and introduced warm-season grasses and native range (McMurphy, 1970; McMurphy et al., 1975; Taliaferro et al., 1975; Rehm, 1984, 1990). In most cases, the reduction in crude protein content resulted from a dilution effect, in which application of P with N fertilizer increased dry matter yields more than N uptake. Several studies reported that P fertilization had no effect on forage digestibility (Taliaferro et al., 1975; Rehm et al., 1972, 1977). Fertilization with P increased tissue P concentrations in native warm-season grasses (Rehm et al., 1977; Kroth & Mattas, 1982; Morris et al., 1982), and luxury consumption of P by switchgrass produced a forage P concentration of 2.2 mg P kg^{-1} (Kroth & Mattas, 1982).

Application of P can produce undesirable changes in the species composition of warm-season grass stands when cool-season grass species are present. In a Nebraska sandhills subirrigated meadow, Brejda et al. (1989) used atrazine [6-chloro-N-ethyl-N'-(1-methylethyl)-1,3,5-triazine-2,4-diamine] to shift the species composition from cool- to warm-season grasses. The second year after application of the herbicide the plots were treated with either 50 kg N ha^{-1} or 50 kg N and 18 P ha^{-1}. The combined application of P and N stimulated greater invasion of cool-season grasses, primarily Kentucky bluegrass, on atrazine treated plots compared with plots receiving only N. Similarly, in a tallgrass prairie site in Kansas, Mader (1956) reported greater invasion of Kentucky bluegrass in plots treated with N and P compared with plots treated only with N.

Potassium

Native warm-season grasses have low K requirements and are able to adequately meet their requirements on low K soils without K fertilization. Several studies reported that K had no effect on forage yield, protein content, or digestibility

Table 12–2. Soil P levels, maximum dry matter yield increase, and P-use efficiency, of P-fertilized native warm-season grasses.

Cultivar	Location	Soil P	Range of fertilizer P inputs	Year	N-rate	Maximum DM yield increase over 0-P control	PUE†	Reference‡
		mg kg^{-1}	kg ha^{-1}		kg ha^{-1}	kg ha^{-1}	kg kg^{-1}	
Big bluestem								
Kaw	OK	--	40	1999–1969	0	80	2	2
					45	410	10	
					90	1740	44	
					180	1210	30	
Switchgrass								
Caddo	OK	--	40	1966–1969	0	(–190)	0	2
					45	470	12	
					90	800	20	
					180	2290	57	
Pathfinder	NE	2.3	45–90	79–82	0	(–700)–600	0–13	7
					45	84–2330	1–30	
					90	2246–4980	25–110	
					134	2440–5390	27–99	
Indiangrass								
	OK	--	40	1966–1969	0	(–210)	0	2
					45	(–80)	0	
					90	690	17	
					180	650	16	
Mixture								
	NE	3	8.5–34	1971–1974	0	(–250)–1030	0–89	5
					45	320–1370	17–106	
					90	560–1830	37–124	
					135	720–2260	53–132	
					180	470–2750	55–160	

(continued on next page)

of native warm-season grasses, even when applied in combination with high rates of N and P (Graves & McMurphy, 1969; Rehm et al., 1972; Friedrich et al., 1977; Smith, 1979; Hall et al., 1982; Kroth & Mattas, 1982; Taylor & Allinson, 1982). Potassium fertilization has been shown to increase forage K concentrations (Smith, 1979; Kroth & Mattas, 1982). In Missouri, luxury consumption of K resulted in a forage K concentration of 15 mg K kg^{-1} DM at the early head stage of development (Kroth & Mattas, 1982). Graves and McMurphy (1969) reported that K fertilization had no effect on species composition within a tallgrass prairie in Oklahoma.

Soil pH and Lime

Several studies indicate that native warm-season grasses can tolerate moderately acid soil conditions (Jung et al., 1988). In West Virginia, switchgrass produced >8 Mg ha^{-1} yr^{-1} on a soil of pH 5.0 (Balasko et al., 1984), and in southeast-

Table 12–2. Continued.

Cultivar	Location	Soil P	Range of fertilizer P inputs	Year	N-rate	Maximum DM yield increase over 0-P control	PUE†	Reference‡
		mg kg⁻¹	kg ha⁻¹		kg ha⁻¹	kg ha⁻¹	kg kg⁻¹	
	NE	1.5	22–45	1967	90	720–760	17–32	4
				1968	90	110–340	5–8	
				1969	90	1300–1610	29–72	
				1970	90	360–1100	16–25	
	NE	1	6–24	1974–1979	0	880–1200	25–147	6
					34	170–1230	28–51	
					68	370–1970	62–82	
					101	590–2730	98–114	
					134	770–3450	128–144	
Native range								
	KS	7–50	22	1963	0	(−110)–70	0–3	3
					37	270–810	12–36	
					75	(−850)–670	0–30	
Native meadow								
	OK	--	45	1965	0	(−235)	0	1
					45	600	14	
				1966	0	(−30)	0	
					45	630	14	
				1967	0	80	2	
					45	440	10	
				1968	0	210	5	
					45	1000	22	
				1969	0	190	4	
					45	1510	34	

† P use efficiency calculated as [(forage yield$_F$ − forage yield$_c$)/(fertilizer P applied)] where F = phosphorus fertilized plants and C = unfertilized control.
‡ References: (1) McMurphy, 1970; (2) McMurphy et al., 1975; (3) Moser and Anderson, 1964; (4) Rehm et al., 1972; (5) Rehm et al., 1976; (6) Rehm, 1984; (7) Rehm, 1990.

ern Missouri, indiangrass produced up to 8 Mg ha⁻¹ yr⁻¹ with application of 168 kg N ha⁻¹ on a soil of pH 5.57 (Brejda et al., 1995). In a pot study using five soils representative of the New England region and varying in pH from 4.5 to 5.5, Taylor and Allinson (1982) reported that application of lime alone had no effect on yields of big bluestem, indiangrass, and switchgrass. In Pennsylvania, Jung et al. (1988) seeded switchgrass, big bluestem, and orchardgrass (*Dactylis glomerata* L.) on an acid soil that ranged in pH from 4.3 to 4.9. They reported that the orchardgrass seedlings emerged, turned black, and died. In contrast, switchgrass and big bluestem seedlings were able to persist and establish, although growth was slow. Following establishment, application of 4.5 Mg ha⁻¹ lime increased switchgrass and big bluestem forage yields 0.4 to 1.0 Mg ha⁻¹ yr⁻¹ for a 3-yr period.

Application of lime may increase warm-season grass response to N. In a greenhouse pot study, Taylor and Allinson (1982) reported that yields of big bluestem, switchgrass, and indiangrass did not increase with application of 45 kg N ha⁻¹ when applied without lime, but increased more than three-fold when treated with the same N rate in combination with lime and P. On an acid (pH 4.3 to 4.9) soil in Pennsylvania, Jung et al. (1988) reported a significant lime × N interaction in which yields of switchgrass and big bluestem were greater when N was applied in combination with lime compared with N alone.

PHOSPHORUS, POTASSIUM, AND LIME RECOMMENDATIONS FOR WARM-SEASON GRASSES

The decision whether to apply lime, P, or K fertilizer for establishment of warm-season grasses dependents on the nutrient level, supplying capacity, and pH of the soil, the potential for weed invasion, economic return from application of the nutrient, and environmental concerns. In Missouri, if soil pH is <5.8, application of lime is recommended at the time of seeding. If the soil tests very low or low in P such that P application is warranted, deficiencies are best corrected at seeding when fertilizer may be incorporated into the soil (Rehm, 1990).

Phosphorus recommendations are based upon soil P levels (Table 12–3). Nebraska recommends periodic soil testing and application of P at a rate based on the soil test level (Table 12–3). These P rates have been determined to provide an economic response to P fertilization for a soil test level. In contrast, Iowa, Kansas, and Missouri, subscribe to the concept that to meet the nutritional needs of the plant, the soil P level must be increased to a target level, above which no plant response is expected. This target level is also called sufficiency level or critical level, and is defined as "the minimum soil test level required to produce optimum crop yields" (Brown, 1996). There is some variation within this region concerning what is the appropriate sufficiency level because of differences in soil characteristics (Table 12–3). The behavior of applied P in soils differing in texture, organic matter content, and mineralogy of the clay fraction lead to different rates of mineralization and P fixation (Brown, 1996).

Phosphorus recommendations can vary depending on whether the warm-season grass stands are used for hay production or grazing. With grazing, most of the P and K is recycled in the pasture in livestock excreta. Grazing, however, can result in increased heterogeneity in the distribution of P and K in the pasture, with greater accumulation where livestock congregate, such as watering and resting areas (West et al., 1989; Mathews et al., 1994a). Because of increased accumulation of P and K in areas where livestock congregate, these areas should be avoided or sampled separately when determining fertility recommendations for a pasture (West et al., 1989; Mathews et al., 1994b). In general, P application is needed only if the soil P level falls in the range where an economic response can be expected from fertilization, or below the sufficiency level such that it begins limiting plant growth.

Because more forage is removed when land is hayed, higher fertility recommendations are given for hayland. In Missouri and Iowa, P recommendations are made to replace the P removed in the hay. The application rate will vary with the forage yield. In Missouri, the algorithm for maintenance application of P (kg P_2O_5 ha^{-1}) with warm-season grasses is 1.0 times the yield goal (Mg ha^{-1}). If the soil test level is already in the optimum category (Table 12–3), Iowa recommends application of 6 kg P_2O_5 per Mg of forage removed (Voss et al., 1996).

Table 12–3. Phosphorus recommendations for warm-season grasses in Iowa, Kansas, Missouri, and Nebraska.

State	Soil test levels†	Fertility rating	Recommended application
	mg P kg^{-1}		kg P$_2$O$_5$ ha^{-1}
Iowa‡	0–8	very low	67
	9–15	low	50
	16–20	optimum	33
	21–30	high	0
	>30	very high	0
Kansas§	<5	very low	22
	6–12	low	22
	13–25	medium	11
	26–50	high	0
	>50	very high	0
Missouri¶	0–7	very low	35–80
	8–11	low	20–35
	12–15	medium	0–20
Nebraska#	0–5	very low	45
	6–15	low	22
	16–25	medium	11
	>25	high	0

† Soil test levels are for Bray and Kurtz #1.
‡ Iowa P recommendations vary depending upon subsoil P level. The above recommendations are for low subsoil P levels (Voss et al., 1996).
§ Kansas P recommendations are for soils with a pH < 6.0. For soils with pH between 6.0 and 8.0, no P is recommended for warm-season grasses (Ray Lamond, Kansas State University, 1996 personal communication).
¶ Annual recommendation assuming a 6.7 kg ha^{-1} dry matter yield and an 8-yr build-up treatment (James R. Brown, University of Missouri, 1996 personal communication).
# Anderson and Shapiro (1990).

In addition to a soil test, data from Missouri with switchgrass suggests that a forage test can be used to determine whether P should be applied. If switchgrass forage harvested at the early head stage contains <1 mg P kg^{-1} of forage, P is limiting plant growth and P fertilizer should be applied (Kroth & Mattas, 1982).

Nebraska, Kansas, and Missouri do not recommend K application for warm-season grasses. This seems logical given the absence of a significant yield response to K fertilization. Iowa, however, does give a K recommendation for warm-season grasses which is based on soil test K levels. If the soil contains <60 mg kg^{-1} ammonium acetate extractable K and subsoil K levels also are low, Iowa recommends application of 90 kg K$_2$O ha^{-1}. If the soil contains between 61 and 90 mg K kg^{-1}, application of 67 kg K$_2$O ha^{-1} is recommended. Iowa does not recommend application of K if the soil contains >90 mg K kg^{-1}. If the soil is in the optimum category (91 to 130 mg K kg^{-1}), Iowa also recommends application of 33 kg K$_2$O per Mg of forage removed.

Currently, there is little information on the P, K, or lime requirements of native warm-season grass–legume mixtures. Nebraska makes the general recommendation that if legumes comprise one-fourth or more of the stand, apply 50% more P than for grass alone (Anderson & Shapiro, 1990). The recommendation does not distinquish between warm- and cool-season grass–legume mixtures. In general, legumes have higher P and K requirements and are more sensitive to soil pH than native warm-season grasses. Legume establishment and persistence in warm-season grass stands may be hindered if soil pH or P and K levels are insufficient for good legume growth.

SUMMARY

Proper fertilization can be used to improve stand establishment and increasing forage production and quality of native warm-season grasses; however, fertilization at the wrong time or with the wrong nutrients is economically inefficient, can cause stand degradation through invasion of undesirable species, and may cause environmental degradation through contamination of surface and groundwater.

Native warm-season grass responses to N fertilization are affected by a host of environmental and management factors including precipitation amounts and distribution during the growing season and between years, inherent soil fertility, the species and cultivar, N source, N timing, burning, haying, and grazing. The challenge agronomists and researchers face is to understand and control the multitude of environmental and management factors that influence warm-season grasses to consistently achieve the most efficient use of N fertilizer.

Cool-season legumes interseeded into seeded warm-season grass stands have been shown to increase both forage yield and quality. In some cases this resulted from additional forage production by the legumes; however, some introduced legumes also supplied fixed N to the grasses. Care must be taken in using introduced cool-season legumes with warm-season grasses because several legume species were highly competitive and replaced a portion of the warm-season grass stand. Nitrogen fertilization of native warm-season grass stands may help maintain grass competitiveness; however, there is currently no information on appropriate N rates and timing for fertilization of native warm-season grasses interseeded with highly competitive cool-season legumes.

Recommended N fertilization rates for native warm-season grasses vary from state-to-state and range between 67 to 110 kg N ha^{-1} yr^{-1} for seeded stands, and 33 kg N ha^{-1} yr^{-1} for native range in Kansas. The economic return to fertilization should be calculated before N fertilizer is applied because in some situations it may be cheaper to purchase additional forage or rent pasture than to fertilize. Nitrogen fertilization will not overcome improper management or overgrazing of native warm-season grasses. Appropriate management is as necessary on seeded warm-season grass stands and prairies as on cropland.

Phosphorus fertilization often produced little or no response in native warm-season grasses when applied in combination with low rates of N; however, at N rates of 90 kg N ha^{-1} or greater, P-use efficiency by warm-season grasses may exceed N-use efficiency. Application of P, however, also can promote invasion of cool-season grasses into native warm-season grass stands. Native warm-season grasses have low K requirements and are able to adequately meet their requirements on low K soils without K fertilization. Because cool-season legumes have higher P, K, and pH requirements than native warm-season grasses, their establishment and persistence in warm-season grasses stands may be hindered if soil P and K levels or pH are insufficient for good legume growth. There is, however, currently no information on P, K, or pH requirements of native warm-season grass–legume mixtures.

REFERENCES

Anderson, B.E., and C.A. Shapiro. 1990. Fertilizing grass pastures and haylands. Univ. of Nebraska NebGuide G78-406. Univ. of Nebraska, Lincoln.

Balasko, J.A., D.M. Burner, and W.V. Thayne. 1984. Yield and quality of switchgrass grown without soil amendments. Agron. J. 76:204–208.

Bashan, Y., and H. Levanony. 1990. Current status of *Azospirillum* inoculation technology: *Azospirillum* as a challenge to agriculture. Can. J. Microbiol. 36:591–607.

Becker, D.A., and J.J. Crockett. 1976. Nitrogen fixation in some prairie legumes. Am. Midl. Nat. 96:133–143.

Black, C.C. 1971. Ecological implications of dividing plants into groups with distinct photosynthetic production capabilities. Adv. Ecol. Res. 7:87–114.

Blanchett, K.M., J.R. George, R.M. Gettle, D.R. Buxton, and K.J. Moore. 1995. Establishment and persistence of legumes interseeded into switchgrass. Agron. J. 87:935–941.

Bohn, H.L., B.L. McNeal, and G.A. O'Connor. 1979. Soil chemistry. John Wiley & Sons, New York.

Brejda, J.J. 1996. Evaluation of arbuscular mycorrhiza populations for enhancing switchgrass yield and nutrient uptake. Ph.D. diss. Univ. of Nebraska, Lincoln.

Brejda, J.J., J.R. Brown, and C.L. Hoenshell. 1995. Indiangrass and caucasian bluestem responses to different nitrogen sources and rates in the Ozarks. J. Range Manage. 48:172–180.

Brejda, J.J., J.R. Brown, T.E. Lorenz, J. Henry, J.L. Reid, and S.R. Lowry. 1996. Eastern gamagrass responses to different harvest intervals and nitrogen rates in northern Missouri. J. Prod. Agric. 9:130–135.

Brejda, J.J., R.J. Kremer, and J.R. Brown. 1994. Indications of associative nitrogen fixation in eastern gamagrass. J. Range Manage. 47:192–196.

Brejda, J.J., L.E. Moser, S.S. Waller, S.R. Lowry, P.E. Reece, and J.T. Nichols. 1989. Atrazine and fertilizer effects on Sandhills subirrigated meadow. J. Range Manage. 42:104–108.

Brejda, J.J., D.H. Yocom, L.E. Moser, and S.S. Waller. 1993. Dependence of 3 Nebraska sandhills warm-season grasses on vesicular-arbuscular mycorrhizae. J. Range Manage. 46:14–20.

Briggs, J.M., and A.K. Knapp. 1995. Interannual variability in primary production in tallgrass prairie: Climate, soil moisture, topographic position, and fire as determinants of aboveground biomass. Am. J. Bot. 82:1024–1030.

Brown, R.H. 1978. A difference in N use efficiency in C_3 and C_4 plants and its implication in adaption and evolution. Crop Sci. 18:93–97.

Brown, R.H. 1985. Growth of C_3 and C_4 grasses under low N levels. Crop Sci. 25:954–957.

Brown, J.R. 1996. Fertility management of harvested forages in the northern states, p. 93–112. *In* R.E. Joost and C.A. Roberts (ed.) Nutrient cycling in forage systems. Potash and Phosphate Inst., Manhattan, KS.

Clubine, S. 1986. Multiple uses for native warm-season grasses, p. 10–11. *In* Warm-season grasses: Balancing forage programs in the northeast and southern corn belt. Soil Conserv. Soc., Ankeny, IA.

Coleman, D.C., C.P.P. Reid, and C.V. Cole. 1983. Biological strategies of nutrient cycling in soil systems. Adv. Ecol. Res. 13:1–55.

Craswell, E.T., and D.C. Godwin. 1984. The efficiency of nitrogen fertilizers applied to cereals in different climates. p. 1–55. *In* P.B. Tinker and A. Lauchli (ed.) Advances in plant nutrition. Vol. 1. Praeger Publ. Co., New York.

Cuomo, G.J., and B.E. Anderson. 1996. Nitrogen fertilization and burning effects on rumen protein degradation and nutritive value of native grasses. Agron. J. 88:439–332.

Cuomo, G.J., B.E. Anderson, L.J. Young, and W.W. Wilhelm. 1996. Harvest frequency and burning effects on monocultures of 3 warm-season grasses. J. Range Manage. 49:157–162.

Dahlman, R.C., and C.L. Kucera. 1965. Root productivity and turnover in native prairie. Ecol. 46:84–89.

DuBois, J.D., and L.A. Kapustka. 1983. Biological nitrogen influx in an Ohio relict prairie. Am. J. Bot. 70:8–16.

Elliott, E.T., R.V. Anderson, D.C. Coleman, and C.V. Cole. 1980. Habitable pore space and microbial trophic interactions. Oikos 35:327–335.

Friedrich, J.W., D. Smith, and L.E. Schrader. 1977. Herbage yield and chemical composition of switchgrass as affected by N, S, and K fertilization. Agron. J. 69:30–32.

Garcia, F.O., and C.W. Rice. 1994. Microbial biomass dynamics in tallgrass prairie. Soil Sci. Soc. Am. J. 58:816–823.

Gay, C.W., and D.D. Dwyer. 1965. Effects of one year's nitrogen fertilization on native vegetation under clipping and burning. J. Range Manage. 18:273–277.

George, J.R., K.M. Blanchet, R.M. Gettle, D.R. Buxton, and K.J. Moore. 1995. Yield and botanical composition of legume-interseeded vs. nitrogen-fertilized switchgrass. Agron. J. 87:1147–1153.

Giller, K.E., and K.J. Wilson. 1991. Nitrogen fixation in tropical cropping systems. CAB Int., Wallingford, England.

Graves, J.E., and W.E. McMurphy. 1969. Burning and fertilization for range improvement in central Oklahoma. J. Range Manage. 22:165–168.

Hall, K.E., J.R. George, and R.R. Riedl. 1982. Herbage dry matter yields of switchgrass, big bluestem, and indiangrass with N fertilization. Agron. J. 74:47–51.

Hargrove, W.L. 1988. Evaluation of ammonia volatilization in the field. J. Prod. Agric. 1:104–111.

Hayes, D.C., and R.R. Seastedt. 1987. Root dynamics of tallgrass prairie in wet and dry years. Can. J. Bot. 65:787–791.

Hetrick, B.A.D., D.G. Kitt, and G.T. Wilson. 1988. Mycorrhizal dependence and growth habit of warm-season and cool-season tallgrass prairie plants. Can. J. Bot. 66:1376–1380.

Hetrick, B.A.D., G.W.T. Wilson, and J.F. Leslie. 1991. Root architecture of warm- and cool-season grasses: Relationship to mycorrhizal dependence. Can. J. Bot. 69:112–118.

Hobbs, N.T., D.S. Schimel, C.E. Owensby, and D.S. Ojima. 1991. Fire and grazing in the tallgrass prairie: Contingent effects on nitrogen budgets. Ecology 72:1374–1382.

Jung, G.A., J.A. Shaffer, and W.L. Stout. 1988. Switchgrass and big bluestem responses to amendments on strongly acid soil. Agron. J. 80:669–676.

Jung, G.A., J.A. Shaffer, W.L. Stout, and M.T. Panciera. 1990. Warm-season grass diversity in yield, plant morphology, and nitrogen concentration and removal in northeastern USA. Agron. J. 82:21–26.

Kapustka, L.A., and E.L. Rice. 1978. Symbiotic and asymbiotic N_2-fixation in a tall grass prairie. Soil Biol. Biochem. 10:553–554.

Killham, K. 1994. Soil ecology. Cambridge Univ. Press, Cambridge, England.

Klucas, R.V. 1991. Associative nitrogen fixation in plants. p. 187–198. *In* M. Dilworth and A. Glenn (ed.) Biology and biochemistry of nitrogen fixation. Elsevier Sci. Publ. B.V., Amsterdam.

Kneebone, W.R. 1959. An evaluation of legumes for western Oklahoma rangelands. Oklahoma Agric. Exp. Stn. Bull. B-539. Oklahoma State Univ., Stillwater.

Kroth, E.M., and R. Mattas. 1982. Topdressing nitrogen, phosphorus and potassium on warm season grasses for pasture production. Missouri Agric. Exp. Stn. Res. Bull. 1046. Univ. of Missouri, Columbia.

Mader, E.L. 1956. The influence of certain fertilizer treatments on the native vegetation of Kansas prairie. Ph.D. diss. Univ. of Nebraska, Lincoln.

Mathews, B.W., L.E. Sollenberger, V.D. Nair, and C.R. Staples. 1994a. Impact of grazing management on soil nitrogen, phosphorus, potassium, and sulfur distribution. J. Environ. Qual. 23:1006–1013.

Mathews, B.W., L.E. Sollenberger, P. Nkedi-Kizza, L.A. Gaston, and H.D. Hornsby. 1994b. Soil sampling procedures for monitoring potassium, distribution in grazed pastures. Agron. J. 86:121–126.

McMurphy, W.E. 1970. Fertilization and deferment of a native hay meadow in north central Oklahoma. Oklahoma Agric. Exp. Stn. Bull. B-678. Oklahoma State Univ., Stillwater.

McMurphy, W.E., C.E. Denman, and B.B. Tucker. 1975. Fertilization of native grass and weeping lovegrass. Agron. J. 67:233–236.

Morris, R.J., R.H. Fox, and G.A. Jung. 1982. Growth, P uptake, and quality of warm and cool-season grasses on a low available P soil. Agron. J. 74:125–129.

Morris, D.R., D.A. Zuberer, and R.W. Weaver. 1985. Nitrogen fixation by intact grass-soil cores using ^{15}N and acetylene reduction. Soil Biol. Biochem. 17:87–91.

Moser, L.E., and K.L. Anderson. 1964. Nitrogen and phosphorus fertilization of bluestem range. Trans. Kansas Acad. Sci. 67:613–616.

Moser, L.E., and K.P. Vogel. 1995. Switchgrass, big bluestem, and indiangrass. p. 409–420. *In* R.F Barnes (ed.) Forages: An introduction to grassland agriculture. 5th ed. Iowa State Univ. Press, Ames.

Okon, Y. 1985. *Azospirillum* as a potential inoculant for agriculture. Trends Biotechnol. 3:223–228.

Oswalt, D.L., A.R. Bertrand, and M.R. Teel. 1959. Influence of nitrogen fertilization and clipping on grass roots. Soil Sci. Soc. Am. Proc. 23:228–230.

Owensby, C.E., R.M. Hyde, and K.L. Anderson. 1970. Effects of clipping and supplemental nitrogen and water on loamy upland bluestem range. J. Range Manage. 23:341–346.

Owensby, C.E., and E.F. Smith. 1979. Fertilizing and burning Flint Hills bluestem. J. Range Manage. 32:254–258.

Perry, L.J., and D.D. Baltensperger. 1979. Leaf and stem yields and forage quality of three N-fertilized warm-season grasses. Agron. J. 71:355–358.

Posler, G.L., A.W. Lenssen, and G.L. Fine. 1993. Forage yield, quality, compatibility, and persistence of warm-season grass-legume mixtures. Agron. J. 85:554–560.

Power, J.F. 1968. Mineralization of nitrogen in grass roots. Soil Sci. Soc. Am. Proc. 32:673–674.

Power, J.F. 1985. Nitrogen- and water-use efficiency of several cool-season grasses receiving ammonium nitrate for 9 years. Agron. J. 77:189–192.

Rehm, G.W. 1984. Yield and quality of a warm-season grass mixture treated with N, P, and atrazine. Agron. J. 76:731–734.

Rehm, G.W. 1990. Importance of nitrogen and phosphorus for production of grasses established with no-till and conventional planting systems. J. Prod. Agric. 3:333–336.

Rehm, G.W., W.J. Moline, and E.J. Schwartz. 1972. Response of a seeded mixture of warm-season prairie grasses to fertilization. J. Range Manage. 25:452–456.

Rehm, G.W., R.C. Sorensen, and W.J. Moline. 1976. Time and rate of fertilizer application for seeded warm-season and bluegrass pastures: I. Yield and botanical composition. Agron J. 68:559–564.

Rehm, G.W., R.C. Sorensen, and W.J. Moline. 1977. Time and rate of fertilizer application for seeded warm-season and bluegrass pastures: II. Quality and nutrient content. Agron J. 69:955–961.

Risser, P.G., and W.J. Parton. 1982. Ecosystem analysis of the tallgrass prairie: Nitrogen cycle. Ecol. 63:1342–1351.

Sala, O.E., W.J. Parton, L.A. Joyce, and W.K. Lauenroth. 1988. Primary production of the central grassland region of the United States. Ecol. 69:40–45.

Seastedt, T.R. 1988. Mass, nitrogen, and phosphorus dynamics in foliage and root detritus of tallgrass prairie. Ecol. 69:59–65.

Seastedt, T.R., J.M. Briggs, and D.J. Gibson. 1991. Controls of nitrogen limitation in tallgrass prairie. Oecologia 87:72–79.

Sharpley, A.N., and L.W. Reed. 1982. Effect of environmental stress on the growth and amount and forms of phosphorus in plants. Agron. J. 74:19–22.

Sloger, C., and P. van Berkum. 1991. Approaches for enhancing nitrogen fixation in cereal crops. p. 229–234. *In* S.K. Dutta and C. Sloger (ed.) Biological nitrogen fixation associated with rice production. Howard Univ. Press, Washington, DC.

Smith, D. 1979. Fertilization of switchgrass in the greenhouse with various levels of N and K. Agron. J. 71:149–150.

Staley, T.E., W.L. Stout, and G.A. Jung. 1991. Nitrogen use by tall fescue and switchgrass on acidic soils of varying water holding capacity. Agron. J. 83:732–738.

Stanton, N.L. 1988. The underground in grasslands. Annu. Rev. Ecol. Syst. 19:573–589.

Stout, W.L., and G.A. Jung. 1995. Biomass and nitrogen accumulation in switchgrass: Effects of soil and environment. Agron. J. 87:663–669.

Stout, W.L., and G.A. Jung, J.A. Shaffer, and R. Estepp. 1986. Soil water conditions and yield of tall fescue, switchgrass, and caucasian bluestem in the Appalachian northeast. J. Soil Water Conserv. 41:184–186.

Sumner, M.E. 1990. Crop responses to *Azospirillum* inoculation. Adv. Soil Sci. 12:53–123.

Svejcar, T.J., and J.A. Browning. 1988. Growth and gas exchange of *Andropogon gerardii* as influenced by burning. J. Range Manage. 41:239–244.

Taliaferro, C.M., F.P. Horn, B.B. Tucker, R. Totusek, and R.D. Morrison. 1975. Performance of three warm-season perennial grasses and a native range mixture as influenced by N and P fertilization. Agron. J. 67:289–292.

Taylor, R.W., and D.W. Allinson. 1982. Response of three warm-season grasses to varying fertility levels on five soils. Can. J. Plant Sci. 62:657–665.

Tjepkema, J.D. 1975. Nitrogenase activity in the rhizosphere of *Panicum virgatum*. Soil Biol. Biochem. 7:179–180.

Tjepkema, J.D., and R.H. Burris. 1976. Nitrogenase activity associated with some Wisconsin prairie grasses. Plant Soil 45:81–94.

Vlassak, K., E.A. Paul, and R.E. Harris. 1973. Assessment of biological nitrogen fixation in grassland and associated sites. Plant Soil 38:637–649.

Voss, R.D., A.P. Mallarino, and R. Killhorn. 1996. General guide for crop nutrient recommendations in Iowa. Iowa State Univ. Exten. Bull. Pm-1688. Iowa State Univ., Ames.

Warnes, D.D., and L.C. Newell. 1969. Establishment and yield responses of warm-season grass strains to fertilization. J. Range Manage. 22:235–240.

West, C.P., A.P. Mallarino, W.F. Wedin, and D.B. Marx. 1989. Spatial variability of soil chemical properties in grazed pastures. Soil. Sci. Soc. Am. J. 53:784–789.

Westerman, R.L., R.J. O'Hanlon, G.L. Fox, and D.L. Minter. 1983. Nitrogen efficiency in bermudagrass production. Soil Sci. Soc. Am. J. 47:810–817.

Wilson, J.R. 1975. Comparative response to nitrogen deficiency of a tropical and temperate grass in the interrelation between photosynthesis, growth, and the accumulation of non-structural carbohydrate. Neth. J. Agric. Sci. 23:104–112.

Wilson, J.R., and R.H. Brown. 1983. Nitrogen response of *Panicum* species differing in CO_2 fixation pathways: I. Growth analysis and carbohydrate accumulation. Crop Sci. 23:1148–1153.

Wuenscher, M.L., and G.C. Gerloff. 1971. Growth of *Andropogon scoparium* (little bluestem) in phosphorus deficient soils. New Phytol. 70:1035–1042.